Melting & Casting ALUMINIUM

ROBERT J. ANDERSON, B.Sc., Met.E.
CONSULTING METALLURGICAL ENGINEER

*Formerly Metallurgical Engineer, United States Bureau of Mines;
Lecturer in Metallography, Carnegie Institute of Technology;
Research Metallurgist, Bureau of Aircraft Production;
Instructor in Metallurgy, Missouri School of Mines, etc.*

Liberty 12-cylinder aviation engine.

Lost Technology Series
Reprinted by Lindsay Publications Inc.

Melting & Casting
Aluminum
by Robert J. Anderson, B.Sc., Met. E.

Copyright 1987 by Lindsay Publications, Inc., Bradley IL 60915
Chapters reprinted from "The Metallurgy of Aluminum and Aluminum Alloys", originally copyrighted in 1925 by Henry Carey Baird & Co., Inc. Published by Henry Carey Baird & Co., Inc., New York, 1925

All rights reserved. No part of this book may be reproduced in any form or by any means without written permission from the publisher.

ISBN 0-917914-59-7

3 4 5 6 7 8 9 0

WARNING

Remember that the materials and methods described here are from another era. Workers were less safety conscious then, and some methods may be downright dangerous. Be careful! Use good solid judgement in your work, and think ahead. Lindsay Publications, Inc. has not tested these methods and materials and does not endorse them. Our job is merely to pass along to you information from another era. Safety is your responsibility.

Write for a catalog or other unusual books available from:

    Lindsay Publications, Inc.
    PO Box 12
    Bradley, IL 60915-0012

CHAPTER IX.

ALUMINIUM AND ALUMINIUM-ALLOY MELTING PRACTICE.

ALUMINIUM and aluminium alloys are melted in various types of furnaces, and there is really no standardized mode of melting. In foundry practice, light aluminium alloys have been melted in all types of furnaces that have been used for brass and bronze. The iron-pot furnace, so widely used for melting so-called white metals, is the only one employed for aluminium alloys that is not used for brass and bronze. Actually, a great variety of furnaces are in use for melting aluminium and aluminium alloys,[46] but the iron-pot furnace is generally preferred for melting aluminium alloys in foundry practice and the reverberatory-type furnace for aluminium in rolling-mill practice.[25] Generally speaking, the type of furnace used depends upon whether aluminium or aluminium alloys are melted, and upon the use to which the melted material is put.

One of the principal difficulties in aluminium and aluminium-alloy melting is the prevention of oxidation (dross) losses, and even with the best practice there is always appreciable loss. The fact that the oxidation loss on melting the metal and its alloys is large is well known, but it is doubtful that the factors governing the most suitable methods of melting are equally well known. Melting losses are a serious source of financial loss in aluminium metallurgy because of the high price of the metal. Some idea of the monetary losses incurred in the aluminium industry of the United States owing to oxidation losses on melting may be gained from figures gathered by the author,[33, 38, 39] and reported to the U. S. Bureau of Mines. On the basis of data furnished by operating companies, the gross loss of metal resulting on melting substantially pure aluminium is 2 per cent on the average, and extreme figures of 5 and 0.75 per cent have been reported. The average net loss may be taken as 1.25 per cent, allowing for the recovery of mechanically entangled metal from the dross, and considering dross to contain 40 per cent metallics. The gross loss of metal on melting aluminium alloys in foundry practice may be taken as 4 per cent on the basis of reported figures, and extremes of 8 and 1 per cent have been given. This refers to melting in a variety of furnaces, with proper account taken of the relative tonnage melted therein. The average net loss may be taken as 2.5 per cent. In addition to oxidation losses in melting, there are heavy fuel losses owing to poor operation and badly designed furnaces. The fuel consumption in melting aluminium and aluminium alloys is about twice as great as in melting copper and its alloys. Figures for the monetary losses incurred in melting practice in aluminium metallurgy have been derived by the author [33] as applying to the United States. On the basis of 250,000,000 lb. of aluminium melted, the net loss is about 9,000,000 lb. of metal, assuming that all the metallics in the dross are recovered. With aluminium at 30 cents per lb., the monetary loss due to oxidation is $2,700,000; and setting aside additional losses due to fuel inefficiency, if the dross loss could be reduced 50 per cent, substantial savings would accrue. From a monetary point of view, the study of melting practice for aluminium and its light alloys is of importance, and it is also important from the metallurgical point of view, since faulty melting seriously impairs the properties of the resultant manufactures.

The fundamental requirements for melting are these: (1) the melting losses should be low; (2) the melting should be rapid; and (3) the cost of melting should be low. It is rarely possible that these requirements are met in any furnace selected at random, and it follows that the design and operation of furnaces, particularly for melting an expensive metal like aluminium, warrant the most extensive study and tests. Taken by and large, not a great deal of study has been given to the design of furnaces to be used especially for aluminium and its alloys, and the tendency has been simply to employ furnaces that have previously been used for non-ferrous metals and alloys in general. Many statements are made in the literature to the effect that the melting temperature should be low, and that care should be exercised to prevent overheating and burning the metal. It is also stated by some writers that aluminium and its alloys absorb large quantities of gas on melting, and therefore open-flame furnaces should not be employed. These statements are vague, to say the least, and instructions of this character are entirely useless. At the same time, it is common knowledge that good practice calls for heating only to, or a little above, the correct pouring temperature, and this varies depending upon whether aluminium is melted for casting into rolling ingots or whether aluminium alloys are melted for making castings. The pouring temperature for aluminium in casting rolling ingots is preferably 700 to 750° C., while in sand practice with 92 : 8 aluminium-copper alloy the pouring temperature may vary between 700 and 850° C., depending upon the type of casting. The melting temperature should not be more than 50 to 100° C. above the pouring temperature. Most of the factors affecting results in melting practice have been discussed in detail in a series of papers by the author,[15, 25, 38, 39, 46] to which reference may be made. In the present chapter, the more theoretical aspects of melting practice will be treated only briefly, but description of the principal types of furnaces employed in practice and discussion of their operating details will be given more attention.

THEORETICAL AND PRACTICAL ASPECTS OF MELTING.

The efficiency of the average metal-melting furnace is notoriously low, that of fuel-fired brass-melting furnaces being not over 10 per cent, while that of aluminium and aluminium-alloy furnaces is even lower—apparently not more than 5 per cent, on the average. Apparently there is great need for fuel savings in melting aluminium and its alloys, and the brief treatment of the more important factors affecting results, as given below, may be found suggestive.

Heat Required in Melting.—There seems to be considerable uncertainty as to the efficiency of aluminium and aluminium-alloy melting furnaces, but the operating efficiency of these furnaces can be readily determined with accuracy. If the fuel efficiency of any melting furnace is to be determined, it is necessary to know the theoretical and actual amounts of fuel consumed in melting a unit weight of metal. Then assuming a fuel of definite calorific value, the quantity of fuel required can be calculated. The furnace efficiency, in per cent, may then be taken as equal to

$$\frac{\text{Theoretical amount of fuel}}{\text{Actual amount of fuel}} \times 100 \text{ per cent.}$$

Aluminium has a high specific heat and high latent heat of fusion; both values are considerably higher than those for brass and copper and the non-ferrous alloys in general. Hence, despite the fact that the melting point of 70 : 30 brass, for example, is much higher than that of the 92 : 8 aluminium-copper alloy, it requires about twice as much fuel to melt the latter. A few calculations are given below to show the amount of heat required to melt and also to superheat aluminium. No calculations are given for the heat required to melt the various light aluminium alloys, because the fundamental physical constants necessary have not been determined, so far as is known.

If it is desired to determine the amount of heat required to melt aluminium, the general equation

$$W = t \times Sm(0 - t) + R$$

may be used, where

W = total heat required in gm.-cals.,

t = melting point in ° C. from zero,

$Sm(0 - t)$ = mean specific heat between 0° C. and t° C., and

R = latent heat of fusion in gm.-cals.

Then using the following values for substantially pure aluminium,

$t = 658.7°$ C.,

$Sm(0 - t) = 0.2220 + 0.00005t$, and

$R = 76.8$,

and substituting in the formula,

$$W = 658.7 \times (0.2220 + 0.00005 \times 658.7) + 76.8$$

= 244.7 cals. per gm. to raise aluminium from zero to the melting point. Since aluminium is normally at about 25° C. (room temperature) when charged, the amount of heat actually required will be slightly less. The calculations will be altered to conform to this condition, and they will explain incidentally the saving of heat which will result in charging a hot furnace at night with cold metal, thereby allowing absorption of heat. In the morning when the furnace is started, less fuel will be required to melt the charge than would be the case if the furnace were allowed to stand empty over night. In order to find the amount of heat required to raise aluminium from any temperature t' to the melting point, the usual calculation may be made. If it is required to find the amount of heat necessary to raise aluminium from 25° C. to the melting point, then

$$W = 658.7 \times (0.2220 + 0.00005 \times 633.7) + 76.8$$
$$= 243.9 \text{ cals. per gm.}$$

If the metal were at a higher initial charging temperature, less calories would be required. Although it is inconvenient and unscientific to work in unrelated heat units, but since in American combustion practice the British thermal unit instead of the calorie is employed, the amount of heat required to melt a pound of metal will be expressed in the English system rather than in the metric. The B.t.u. per lb. of aluminium required can be obtained by the following:

$$243.9 \times 1.8 = 439 \text{ B.t.u. per lb.}$$

The calculations to be given later will be on the basis of the amount of fuel required to melt a pound of metal.

In casting aluminium ingots in rolling-mill practice, it is normally necessary to superheat the charges to 750 to 800° C., the ingots being poured at 700 to 750° C. If, for the purpose of an example, it is assumed that the metal is to be raised to 800° C., calculation will show the amount of heat required. The formula

$$W = t + Sm(0 - t) + R + (t' - t) \, S \text{ liquid}$$

may be used, where

W = total heat required in gm.-cals.,

t = melting point in ° C. from zero,

$Sm(0 - t)$ = mean specific heat between 0° and $t°$,

R = latent heat of fusion in gm.-cals.,

t' = temperature heated above t, and

S liquid = mean specific heat in the liquid state.

Since, for practical purposes, the information wanted is the amount of heat required to raise aluminium from 25 to 800° C., and remembering that the heat required to raise the metal from 25° to the melting point is 243.9 cals., then it is only necessary to add to this the amount of heat required to raise the metal from 658.7 to 800° C. Substituting in the last part of the formula, $(t'-t)$ S liquid, the figures are

$$(800 - 658.7)0.308 = 43.5 \text{ cals.},$$

and adding this to 243.9, it follows that 287.4 cals. per gm. are required.

$$287.4 \times 1.8 = 517 \text{ B.t.u. per lb.}$$

Apparently, less heat is required to melt the light alloys than to melt aluminium. It will be considered for the purposes of the present chapter that it requires about 245 cals. to raise 1 gm. of aluminium or of the light aluminium alloys to the melting point, and about 287 cals. per gm. or 517 B.t.u. per lb. to melt and superheat to 800° C., as will be required for pouring ingots or sand castings. Referring to the amounts of different fuels required to melt a unit weight of metal, the quantities shown in Table 74, for fuels of different calorific values, are needed to melt and superheat 100 lb. of aluminium to 800° C., on the basis of a furnace operating at 100 per cent efficiency.

TABLE 74.—*Amounts of different fuels required to melt and superheat 100 lb. of aluminium.*

Fuel.	Calorific value in B.t.u.	Amount required to melt and superheat aluminium to 800° C.
Bituminous coal	12,000 per lb.	4.3 lb.
Anthracite	14,000 per lb.	3.7 lb.
Metallurgical coke	13,000 per lb.	4.0 lb.
Natural gas	900 per cu. ft.	57.4 cu. ft.
Illuminating (city) gas	600 per cu. ft.	86.2 cu. ft.
Producer gas	125 per cu. ft.	431.0 cu. ft.
Fuel oil	19,000 per lb.	2.7 lb.

These figures may be compared with those given by Gillett [9] for the fuel required to melt and heat red brass to the pouring temperature, viz., 2 lb. of coal or coke; 1.4 lb. or 0.18 gal. of fuel oil; 26 cu. ft. of natural gas; 41.5 cu. ft. of city gas; and 217 cu. ft. of producer gas. Roughly, it requires about twice as much fuel to melt the light aluminium alloys as it does to melt the brasses and bronzes. If a furnace consumes 30 lb. of coal per 100 lb. of aluminium melted, the efficiency is evidently only about 14.3 per cent. The fuel efficiency of a furnace can thus be readily calculated on the basis of the amount of fuel actually consumed, and the amount of fuel theoretically required.

Oxidation and Dross Losses.—The oxidation of aluminium has been discussed in Chapter IV, and dross losses in melting practice may be considered in the light of the information there given. Metallic aluminium is normally covered with a thin film of aluminium oxide, since the metal oxidizes in ordinary air at room temperature. When heated in air at temperatures up to the melting point, small particles of aluminium gradually go over to an aluminium oxide, despite the supposed protective effect of the surface film. Of course, if an unbroken film of aluminium oxide, with a melting point of about 2,050° C., covers a piece of aluminium, it will offer some protection against the effects of both heat and oxygen. If the film is broken, then oxidation starts at the break. It has been found in furnace practice that the gross melting loss will be less if a layer of dross is maintained on the surface of the bath to protect the metal from contact with oxygen in the furnace atmosphere. It is, therefore, evident that the total oxidation loss will be increased on melting when fresh surfaces of the bath are exposed to the furnace atmosphere; consequently, the bath should be stirred infrequently in order to avoid breaking the surface scum. In all types of furnaces the loss on melting is a function of the temperature, the oxygen content of the furnace, the amount of stirring, and some other factors. Table 75 gives some figures for the gross and net loss on melting 92 : 8 aluminium-copper alloy in different types of furnaces.

As indicated, the melting temperature for aluminium and its alloys should be low and overheating should, in all cases, be avoided. In foundry practice, it is known from experience that overheating and repeated melting have deleterious effects upon the quality of aluminium alloys, but this is disputed in the case of substantially pure aluminium by Rosenhain and Grogan.[41] The results of tests by these investigators are in striking contradiction to the views generally current, since they find that overheating and repeated melting have no deteriorating effect on the tensile properties of aluminium. However, these tests should not be regarded as proof that overheating and repeated melting under foundry and rolling-mill conditions will not have appreciably bad effects upon the quality of aluminium and aluminium-alloy manufactures.

TABLE 75.—*Melting losses for 92 : 8 aluminium-copper alloy in different furnaces.*

Type of furnace.	Fuel used.	Capacity, lb.	Net melting loss, per cent.	Gross melting loss, per cent.
Stationary crucible	Natural gas	50	1.75	2.5
Stationary crucible	Oil and gas	40	....	5.0
Stationary crucible	Oil and gas	90	2.0	6.0
Stationary crucible	Oil	100	1.0	1.0
Pit	Coke	80	....	2.0
Pit	Coke	80	....	2.0
Pit	Coke	110–325	....	3.0
Pit	Coke	300	1.25	2.0
Pit	Coke	50	1.0	2.25
Iron pot	Natural gas	200	....	4.0
Iron pot	Oil and gas	280	2.0	6.0
Iron pot	Oil and gas	100	....	2.0
Iron pot	Oil and gas	100	....	3.0
Iron pot	Oil	280	....	6.0
Iron pot	Oil	300	....	6.0
Iron pot	Oil	200	1.5	2.0
Iron pot	Oil	400	3.0	4.5
Open-flame	Natural gas	160–200	....	8.0
Open-flame	Oil	200	5.0	8.0
Open-flame	Oil	300	....	4.0

The aluminium oxide (dross) formed on aluminium and aluminium-alloy melts may float on top of the bath or sink to the bottom, depending upon conditions. The specific gravity of aluminium at 800° C. is 2.34, while that of its amorphous oxide is 3.85. Hence, from the point of view of specific gravity alone, the dross should sink, and the fact that some of the oxide does sink is shown by reverberatory-furnace practice in melting substantially pure aluminium, where a fairly large accumulation of aluminium oxide builds up on the furnace bottom after a few days of continuous melting. It should be stated that if aluminium oxide is not " wetted " by liquid aluminium, it will float; if it is wetted, the specific gravities and surface tension must be taken into account. Obviously, if the oxide is wetted by aluminium, it will sink to the bottom of the bath, and the explanation of why a surface covering of dross is always found on top of liquid aluminium must be sought in consideration of the effect of surface tension. Both the viscosity of aluminium and its surface tension will assist in maintaining an oxide scum on the surface.

In melting aluminium and its alloys in fuel-fired furnaces, the question of vaporization of the materials is relatively unimportant. The boiling point of aluminium is at least 1,800° C., and this temperature is rarely attained, even under the worst conditions of firing and overheating. Vaporization of the metal may occur, however, in direct-arc electric furnaces. In the case of zinc-bearing aluminium alloys, the question of zinc losses due to vaporization becomes important, and melting temperatures require careful control when running these alloys. The nitridation of aluminium on melting must be taken into account, since aluminium nitride is normally formed by interaction with nitrogen of the air (cf. Chapter IV).

Furnace Fuels.—Various kinds of fuels are burned in aluminium and aluminium-alloy melting furnaces, and the type of fuel selected is governed by the furnace and economic considerations. Bituminous coal is burned in some reverberatory furnaces for melting aluminium in rolling-mill practice, but it is rarely used in other types of furnaces. Anthracite is used to a subordinate extent in pit furnaces in foundry practice, but coke is the preferred fuel for these furnaces. Natural gas is, of course, the most popular gaseous fuel, and it is used wherever available. City gas or illuminating gas is employed only slightly because of its high cost. Neither producer gas nor water gas have been used to any extent, but there is a tendency for larger founders to install gas producers or generators for manufacturing gas for their own consumption. Fuel oil is used extensively for melting in foundry practice where natural gas can not be obtained. Table 76 gives some figures for fuel consumption and melting costs for melting 92 : 8 aluminium-copper in different types of furnaces. Detailed consideration of the reactions that take place when the various metallurgical fuels are burned in relation to melting losses, furnace operation, furnace atmospheres, and to gas occlusion among other items, is of importance in dealing

TABLE 76.—*Fuel consumption in melting 92 : 8 aluminium-copper alloy in different furnaces compared.*

Type of furnace.	Fuel burned.	Amount of fuel used per furnace, per day.	Average heat charged, weight in lb.	Amount of fuel required to melt 1 lb. of alloy.	Amount of fuel required to melt 100 lb. of alloy.	Cost of fuel per lb. of metal melted.[a]	Cost of fuel per 100 lb. of metal melted.[a]	Approximate furnace efficiency, per cent.[b]
Stationary crucible....	Oil	....	100	0.10 gal.	10.0 gals.	$0.0080	$0.80	3.5
Stationary crucible....	Oil	....	60	0.15 gal.	15.0 gals.	0.0120	1.20	2.3
Stationary iron pot....	Oil	15.0 gals.	300	0.0125 gal.	1.25 gals.	0.0010	0.10	28.0
Stationary iron pot....	Oil	23.0 gals.	100	0.05 gal.	5.0 gals.	0.0040	0.40	7.0
Stationary iron pot....	Oil	....	250	0.0375 gal.	3.75 gals.	0.0013	0.13	9.3
Stationary iron pot....	Natural gas	3600 cu. ft.	100	10.0 cu. ft.	1000 cu. ft.	0.0040	0.40	5.7
Tilting iron pot......	Oil	90.0 gals.	200	0.05 gal.	5.0 gals.	0.0040	0.40	7.0
Open-flame pear-shaped.	Natural gas	8000 cu. ft.	200	4.5 cu. ft.	450 cu. ft.	0.0018	0.18	12.7
Open-flame pear-shaped.	Oil	18.0 gals.	300	0.02 gal.	2.0 gals.	0.0016	0.16	17.5
Pit..................	Coke	136 lb.	200	1.7 lb.	170 lb.	0.0055	0.55	2.3
Pit..................	Coke	....	60	1.0 lb.	100 lb.	0.0033	0.33	4.0
Pit..................	Coke	....	50	2.0 lb.	200 lb.	0.0066	0.66	2.0
Pit..................	Coke	400 lb.	80	0.5 lb.	50 lb.	0.0016	0.16	8.0
Pit..................	Coke	....	60	1.0 lb.	100 lb.	0.0033	0.33	4.0
Pit..................	Coke	....	60	1.0 lb.	100 lb.	0.0033	0.33	4.0
Pit..................	Anthracite	....	60	0.75 lb.	75 lb.	0.0030	0.30	5.3

[a] On the basis of the following assumed costs for the different fuels; oil, $0.48 per gal.; natural gas, $0.40 per 1000 cu. ft.; city gas, $0.80 per 1000 cu. ft.; coke, $6.50 per ton; and anthracite $8 per ton.
[b] On the basis of the following approximate quantities of the different fuels required for melting 100 lb. of alloy: oil, 0.35 gal.; natural gas, 57 cu. ft.; city gas, 86 cu. ft.; coke, 4 lb.; and anthracite, 4 lb.

with the melting problem for aluminium and its alloys. These aspects of the subject, together with examination of the various fuels on the basis of calorific power per unit of cost, and their applicability and limitations, have been discussed at length by the author [39] in another place, but owing to the confines of space they can not be given detailed treatment here.

Furnace Atmospheres.—The constitution of the atmospheres obtaining in melting furnaces is of especial interest when considering furnaces for aluminium and aluminium alloys because of the effects of the different constituents of atmospheres upon melting losses and the quality of the metal. The constitution of atmospheres in aluminium-alloy melting furnaces has been examined in a fairly detailed way by the present author [34, 35] and J. H. Capps. In speaking of furnace atmospheres, it is usual, in foundry parlance, to refer to them as "oxidizing," "neutral," or "reducing." In melting aluminium and other non-ferrous metals and alloys, most melters try to maintain a reducing atmosphere and prevent an oxidizing atmosphere in the furnaces. From the metallurgical point of view, in order to determine whether an atmosphere is oxidizing, neutral, or reducing, it is necessary to know (1) the exact chemical composition of that atmosphere; (2) the kind of metal or alloy melted; and (3) the effect of the various constituents of the atmosphere upon the metal or alloy at various temperatures. Without these data an atmosphere can not be defined. Thus, carbon dioxide may be neutral or oxidizing; it is neutral to liquid copper, but oxidizing to zinc. Magnesium is oxidized by carbon dioxide with the formation of magnesium oxide and the deposition of carbon. Nitrogen, so customarily regarded as an inert gas, readily forms aluminium nitride by interaction with aluminium at moderate temperatures. Unfortunately, the interactions of the various gases existing in furnace atmospheres with aluminium at different temperatures have not been thoroughly investigated, and it is consequently speculative to say what are the effects of the gases upon melting losses.

A general axiom in melting practice is that dross losses are higher when the melting is conducted in an oxidizing atmosphere than in a reducing atmosphere, by which is ordinarily meant atmospheres high in oxygen and high in carbon monoxide, respectively. The constitution of any furnace atmosphere is dependent upon a number of governing factors, of which the following are the most important: (1) kind and quality of the fuel used; (2) type and design of the furnace; and (3) operating conditions, including volume of air supply in relation to fuel supply. From the point of view of correct melting practice it is important to consider carefully the various constituents that may occur in furnace atmospheres, and the possible interactions of these constituents with the metal or alloy melted, at high temperatures. The usual constituents found in melting furnace atmospheres include the following: carbon dioxide (CO_2), carbon monoxide (CO), cyanogen (CN), hydrogen (H_2), methane (CH_4), nitrogen (N_2), oxygen (O_2), sulphur dioxide (SO_2), unsaturated hydrocarbons (C_2H_4), and water vapor. The limiting percentages of the constituents found in a number of different furnace atmospheres are given in Table 77, and in all cases reference is to the actual atmosphere in contact with the metal during melting. Detailed lists of analyses of the atmospheres found in various furnaces have been given elsewhere.[35]

Carbon monoxide, carbon dioxide, cyanogen, hydrogen, methane, sulphur dioxide, and unsaturated hydrocarbons apparently do not interact with aluminium at normal melting temperatures, although they are dissolved by aluminium (cf. Gas Occlusion below). Oxygen and nitrogen are the active interacting gases in melting, and water vapor (present in furnaces where steam is used for atomization of oil) interacts with the metal. The atmosphere in a fuel-fired furnace can be controlled within small limits by proper adjustment of the fuel-and-air supply relations, and practically perfect combustion, as shown by high carbon dioxide and low oxygen content of the atmosphere in open-flame furnaces, is entirely practical. In practice, according to gas analyses made of many operating furnaces, the atmospheres are normally oxidizing in fuel-fired furnaces, but in electric furnaces, both granular-resistor and direct-arc types, they are normally reducing, i.e., high in carbon monoxide. That the electric furnace has high carbon-monoxide content in its atmosphere is not necessarily proof that dross losses will be lower than in a fuel-fired furnace with a high oxygen atmosphere, because in the direct-arc furnace the temperature may be so high locally that the carbon monoxide will interact with the aluminium, yielding aluminium oxide and setting free carbon. In general,

TABLE 77.—*Analyses of gases in contact with aluminium alloys in various types of melting furnaces.*[a]

Type of furnace.	Composition of constituents, per cent by volume.							
	Carbon dioxide, CO_2	Carbon monoxide, CO	Hydrogen, H_2	Methane, CH_4	Unsaturated hydrocarbons, C_2H_4, etc.	Nitrogen,[b] N_2	Oxygen, O_2	Cyanogen, CN
Stationary, oil-fired, iron-pot furnace, no cover	0.0– 0.2	Nil	0.0– 0.2	Nil	0.0–0.4	78.9–79.3	20.2–20.8	Nil
Stationary, gas-fired, iron-pot furnace, with cover	3.5– 9.8	0.0– 0.3	0.0– 0.1	Nil	0.0–0.4	80.0–86.6	3.5–16.5	Nil
Stationary, gas-fired, crucible furnace	6.7–10.8	1.9– 6.3	0.7– 6.2	0.3–4.1	0.0–1.3	77.6–85.4	0.0– 4.5	Nil
Stationary, oil-fired, crucible furnace	6.4–14.5	Nil	0.0– 0.4	Nil	0.0–0.5	80.1–83.6	2.0–12.6	Nil
Tilting, coke-fired crucible furnace, natural draft	2.4– 9.2	0.0– 0.4	0.0– 0.2	0.0–0.2	Nil	79.0–79.7	10.5–17.9	Nil
Tilting, coke-fired crucible furnace, forced draft	12.8–19.1	0.0–11.2	0.0– 0.8	0.0–0.1	Nil	74.4–80.9	0.2– 6.8	Nil
Tilting, oil-fired open-flame furnace, egg-shaped Schwartz type	5.9–14.7	0.0–11.0	0.0– 7.1	0.0–2.0	0.0–0.4	73.3–83.4	0.0–12.3	Nil
Stationary, oil-fired, open-flame furnace, cylindrical Schwartz type	12.8–14.6	0.0– 2.9	0.0– 1.7	0.0–0.5	Nil	82.2–84.5	0.1– 2.0	Nil
Open-flame, oil-fired reverberatory furnace	0.1–12.1	0.0– 0.7	0.0– 0.1	0.1–0.4	Nil	79.0–88.4	2.9–20.8	Nil
Indirect-arc rocking, electric furnace	0.0– 8.4	16.0–42.8	0.0–28.6	0.0–3.0	Nil	47.2–82.6	0.0– 7.6	0.0–2.1
Granular-resistor electric furnace	12.4–19.9	0.9–16.6	0.0– 5.4	0.0–0.5	0.0–0.4	64.2–84.5	0.2– 4.9	Nil

[a] Tests by author and J. H. Capps.   [b] Nitrogen, by difference.

with fuel-fired furnaces, it is best to operate so that the atmosphere is high in carbon and low in oxygen; heavy dross losses are thereby avoided and the greatest fuel efficiency is secured.

Gas Occlusion.—The occlusion of gases by aluminium and its light alloys on melting is of interest rather more from the point of view of casting and working than from the point of view of melting losses, but it is closely dependent upon the constitution of the furnace atmosphere. Liquids, in general, dissolve less gas with increasing temperatures, but liquid metals depart from this law and dissolve more gas with increasing temperatures.[3] Hence, the higher aluminium or its light alloys are heated prior to pouring the more gas will be dissolved. According to the available information, the solubility of gases in liquid metals does not appear to obey Henry's law. Desch[2] states that such solutions readily remain supersaturated, and therefore an overheated melt will usually contain more gas at the moment of pouring than a melt which has been merely heated to the correct pouring temperature. So far, not much work has been done on gases in aluminium, but Gwyer[21] states that hydrogen is the predominant gas in the metal. He found that cast aluminium which showed exceptional blistering after rolling into sheet and annealing contained one-tenth to one-third of its own volume of occluded gas, and that the composition of the gas varied exceedingly. For cast aluminium that had given this difficulty it was found that the gas given off on heating varied as to composition as follows: 4 to 8 per cent carbon dioxide by volume; 0.0 to 7.5 per cent carbon monoxide, 1.0 to 24.5 per cent methane, 55 to 81 per cent hydrogen, 1.5 to 4.5 per cent oxygen, and 5 to 11 per cent nitrogen. The fact that the gas evolved from aluminium is high in hydrogen has been confirmed by other investigators.[7] The effect of dissolved or occluded gases upon the occurrence of blowholes in aluminium-alloy castings has been discussed also by the author.[18]

According to Czochralzki,[42] aluminium begins to dissolve gases in appreciable amount only at 900° C. or higher, and he states that nitrogen is the least dissolved and the quantities of gas increase in the following order: carbon monoxide, air, oxygen, sulphur dioxide, carbon dioxide, coal gas, and hydrogen. In the case of nitrogen, part of the gas is dissolved as aluminium nitride, while oxygen is only partly combined as aluminium oxide, and above 1,200° C. much oxygen is dissolved as such and liberated on freezing. Carbon monoxide and carbon dioxide resemble each other closely in their effects and form aluminium carbide. Sulphur dioxide forms some aluminium sulphide at sufficiently high temperature. A furnace that does not bring the products of combustion into contact with the metal melted is regarded as desirable by some foundrymen, but it is evident that if the combustion products are not dissolved appreciably then open-flame furnaces are entirely suitable. In melting practice for aluminium and its alloys it is axiomatic that the air supplied should be burned as completely as possible, rather than the fuel, so that free oxygen, at a high temperature, will not be passed over the metal with the products of combustion. In the light of the available evidence, oxygen, nitrogen, and hydrogen appear to be the most harmful gases—the first two because they combine with aluminium at normal melting temperatures, and hydrogen because it appears to be readily soluble in the metal. Probably any furnace can be operated so as to give but relatively little trouble from dissolved gas, but if a suitable liquid (molten) flux cover is used on the surface of the metal it would appear that some of the present oxidation difficulties could be reduced.

Covers and Fluxes.—Fluxing covers are employed to a slight extent in aluminium metallurgy for the purpose of preventing oxidation on melting. The usual liquid (molten) flux cover consists of a mixture of salts which is liquid at, or preferably below, the normal melting temperature. Thus, a mixture of 60 per cent potassium chloride and 40 per cent cryolite has been employed. Sodium chloride has been used as a liquid cover in melting aluminium in reverberatory furnaces for casting into rolling ingots, and in melting aluminium alloys in foundry practice. Borax, a mixture of sodium chloride and borax, a mixture of sodium chloride and calcium fluoride, and complex mixtures of a number of salts including potassium chloride, sodium chloride, lithium fluoride, sodium carbonate, etc., have been used for the prevention of oxidation on melting. Hill [27] and his co-workers have examined the effect of melting No. 12 alloy under boric oxide for 30 mins., finding that the aluminium oxide content, as determined by Rhodin's method, was reduced from 2.27 to 1.70 per cent by this treatment. These investigators point out that it is conceivable that on melting a metal under a liquid cover, which dissolved metallic oxides, the metal would have its oxide content lowered by solution in the covering, which would in turn cause a diffusion of the oxide from the interior portions richer in oxide. Thus, in time, an appreciable quantity of oxide might be dissolved in the molten cover. Although liquid covers are used more frequently for the purpose of preventing oxidation rather than for dissolving out aluminium oxide, it is evident that such covers fail in their purpose if they are readily penetrable by gases. Hill and his co-workers state that many of the common coverings are readily penetrable by hydrogen. Zinc chloride is the flux ordinarily employed in melting practice for cleansing melts of occluded dross, but the tendency in foundry melting is to do away with fluxes of any kind. The effect of zinc chloride as a flux is largely mechanical, since it volatilizes rapidly when added to a liquid bath, and causes agitation in the bath whereby any suspended dross is carried to the surface. Rosenhain and Archbutt [48] have recently discussed the use of fluxes in melting aluminium and its alloys, they pointing out that fluxes are entirely unnecessary.

Melting Pots and Crucibles.—The question of the material composing the melting pot in pot or crucible furnaces, or of the lining in open-flame furnaces, is of especial importance in aluminium and aluminium-alloy melting because of the great chemical activity of aluminium. Some preliminary work on the effect of heating aluminium in crucibles of different kinds has been carried out by Hill [27] and his co-workers. In these experiments, aluminium was melted in an oxidizing atmosphere in different kinds of crucibles, without stirring, at 750° C. It was shown that, even with prolonged heating on exposure to air, aluminium does not take up much oxide from surface contact with the atmosphere. On prolonged heating in refractories containing even combined silica, reduction takes place with the introduction of alumina. The results are given in Table 78. In foundry practice, light aluminium alloys are melted both in graphite-clay crucibles and in cast-iron melting pots. Steel has been tried for melting pots, but is practically useless for this purpose because the rate of attack of liquid aluminium on steel is rapid.

Gray cast-iron and so-called semi-steel pots are largely used in foundry melting. The life of an iron pot is dependent upon a number of factors, of which the chemical composition appears to be a relatively unimportant one, although the effect of this has not been studied thoroughly. The life of an iron pot is dependent, among other things, upon the rate of dissolution of cast iron in liquid aluminium alloys, upon the quality of the pot, i.e., flaws and other defects in the pot, upon the temperatures employed, and upon whether washes or coatings are used on the inside and outside of the pot. The principal advantages which arise from using iron pots rather than graphite-clay crucibles for melting lie in the fact that larger capacities can be obtained with the former, the speed of melting is more rapid, and a liquid heel

TABLE 78.—*Results of heating aluminium in different kinds of crucibles.*[a]

Crucible.	Al_2O_3, per cent.	Fe, per cent.	Cu, per cent.	Si, per cent.	
				Total.	Graphitic.
Original ingot............	0.35	0.34	0.10	0.30	......
Pressed steel.............	0.40	0.50	0.12	0.30	......
Acheson graphite block.....	0.40	0.29	0.10	0.23	......
Dixon graphite-clay........	0.90	0.34	0.10	0.25	......
Magnesia................	1.30	0.36	0.13	0.30	......
Clay....................	2.70	0.40	0.10	0.94	......
Porcelain (glazed).........	7.50	0.45	0.10	3.00	2.80
Sand....................	11.40	0.30	0.10	6.40	5.80
Fused silica..............	14.70	0.30	0.10	11.30	11.00 [b]

[a] Based on Hill, Thomas, and Vietz.
[b] Crucible turned black in contact with metal.

can be employed. The principal disadvantage results from the fact that iron pots are liable to give rise to "hard spots" in the resultant castings,[40] unless the melting is properly handled. Hard inclusions may be largely overcome by scraping and cleaning the inside surface of the pots so that no accretions build up (cf. Chapter XII). When an aluminium alloy is melted in an iron pot, a hard complex iron-aluminium alloy scale builds up on the inside of the pot; this material may be knocked off into the melt on stirring, and cause so-called hard spots in the castings. Fig. 99 shows the fracture of a piece of hard iron-pot scale from an accretion built up in a stationary iron pot; and Fig. 100 shows inclusions of iron-pot scale in 92 : 8 aluminium-copper alloy.

Aluminium-Alloy Melting Practice

The chemical composition of cast-iron pots is variable, and a foundry company which specializes in the production of pots for aluminium-alloy melting makes pots of the following typical analysis: 0.50 per cent combined carbon, 3.15 total carbon, 0.44 manganese, 0.269 phosphorus, 0.198 sulphur, and 2.57 per cent silicon. Iron pots will last from 8 to 12 days melting 9 to 12 hrs. per day, and most foundrymen prefer to remove old pots after the eighth day rather than run the risk of losing a heat owing to a breakout. Before a new pot is put into service it should be carefully inspected for blows, draws, and other foundry defects, as well as for evidence of patching, since otherwise a faulty pot may

Fig. 99.—*Fracture of hard iron-pot scale from an accretion built up in a stationary iron pot;* ×1.

Fig. 100.—*Inclusions of iron-pot scale in 92 : 8 aluminium-copper alloy; unetched; vertical illumination;* ×7 (cf. Fig, 145).

be placed in the furnace and failure will take place after a few heats. The use of washings and coatings for cast-iron pots for the purpose of prolonging their life is generally recognized as desirable, but the main difficulty comes in securing a suitable wash. Various high-temperature cements, magnesia, fire clay, and other materials have been employed with indifferent success, but it is reported * that white-washing the inside with slaked lime has given good results. The use of coatings on the outside surface of cast-iron pots for the purpose of preventing oxidation and protecting the iron from the cutting action of the flames has been suggested and used by some founders. The main difficulty lies in obtaining a coating which will adhere to the pot. Some high-temperature cements have been tried for this purpose. Iron pots fail in service through defects in the original casting which escape inspection, through stretching and scaling (oxidation), and through breakouts from the inside where the pot becomes thin in spots owing to the attack of the aluminium. In iron-pot melting, the iron content of the resultant alloys is always higher than when melting in crucibles or in contact with a refractory furnace lining, owing to the dissolution of iron by the aluminium. If any iron oxide is present on the surface of the pot, or if iron oxide falls into the aluminium from the flange of the pot, then the reaction

$$Fe_2O_3 + 2Al = Al_2O_3 + 2Fe$$

takes place. The reduced iron alloys with the aluminium and the aluminium oxide is found in the dross. Of course, the rate of attack by liquid aluminium alloys upon cast-iron pots is a function of the composition of the alloy. Zinc-containing aluminium alloys have much greater corrosive action than have aluminium-copper alloys. The question of melting in cast-iron pots has been discussed at length in another place by the author,[38] but owing to the confines of space it is not possible to deal with this at as great length as is desirable and warranted. Table 79 gives some data as to the life of cast-iron pots in stationary furnaces.

A relatively small tonnage of aluminium alloys is melted in graphite-clay crucibles, either in pit or crucible furnaces. Although the composition of crucibles made by different makers may vary considerably, a so-called plumbago crucible of a good standard grade is made of a mixture of about 40 to 55 per cent graphite, 5 per cent silica sand, and the remainder clay of Klingenberg grade or equivalent. The life of graphite-clay crucibles

---
* Private communication, Jan. 5, 1921.

TABLE 79.—*Life of gray cast-iron pots in stationary furnaces.*

Fuel used.	Size of pot, ins. Diameter.	Size of pot, ins. Depth.	Capacity, lb.[a]	Thickness, ins.	Number of heats.	Metal melted before failure, lb.
City gas	15	15	200	0.375	24–50	4,800–10,000
Natural gas	18	18	250	1	40	10,000
Natural gas	$20\frac{1}{8}$	$15\frac{5}{8}$	300	$\frac{3}{4}$ in. side-wall, 1 in. bottom	50–55	15,000–16,500
Blue water gas	36	30	300	$\frac{4}{7}$ [b]	48	14,400
Gas and oil			100	0.75	30–50	2,000–5,000
Gas and oil			100	1	c	12,500
Oil			300	$\frac{7}{8}$	20	6,000
Oil	18	18	250–300	0.75	56–80	14,000–24,000
Oil	d	d	1000–1200	1.0	24–40	24,000–48,000
Oil [e]	22	12	200	1 in. bottom tapered to 0.5 in. at flange.	50	10,000
Oil [e]	30	16	400	0.5 in. sides, $\frac{7}{8}$ in. bottom.	85	34,000

[a] Operating capacity, not actual. [b] Covered on the inside and outside with a high-temperature cement.
[c] Life, about 21 days on continuous melting. [d] Rectangular pot, 16 ins. wide by 37 ins. long by 26 ins. deep.
[e] Tilting iron-pot furnace.

used for melting aluminium alloys may, in general, be expected to be longer than when used for brass and bronze, owing to the lower melting temperatures employed, although the corrosive action of aluminium alloys is greater than that of brass and bronze. In contact with siliceous materials, aluminium will reduce silica at the normal melting temperature according to

$$4Al + 3SiO_2 = 3Si + 2Al_2O_3.$$

The reduced silicon enters the aluminium and aluminium oxide is formed. Gillett [9] has discussed crucible life, as related to brass melting, and the proper treatment of crucibles in the foundry, and the principles elucidated are generally applicable to aluminium-alloy melting.

Refractories and Linings.—In the case of tilting and stationary open-flame furnaces, reverberatory furnaces, and electric furnaces, aluminium and its light alloys are melted in contact with refractory linings and bottoms. The question naturally arises: what is the correct refractory to employ? In this connection there are several factors to be carefully studied, viz., (1) the chemical composition of the material melted; (2) the chemical composition of the refractory and its general properties; (3) interaction of aluminium and the constituents of the refractory; and (4) cost of the refractory. Given adequate data, the most suitable refractory lining for a given installation can be readily determined. The conditions surrounding the life of the refractory lining of the furnace shell in pit furnaces, crucible furnaces, and iron-pot furnaces, as contrasted with the lining in open-flame furnaces, reverberatories, and electric furnaces, are similar, but in the three latter types the refractory must withstand the corrosive action of liquid metal and mechanical abrasion due to charging in addition to high temperatures and rapid changes in temperature.

Generally speaking, refractory fireclay bricks have been used for lining open-flame furnaces, although in recent years there has been a tendency to employ a great variety of materials for this purpose. In open-flame tilting furnaces (Schwartz type), a rammed lining of ganister and fireclay has been employed with considerable success, and the life of this material is long when aluminium alloys are melted. Carborundum fire sand is employed for open-flame tilting and stationary furnaces, and for the hearth in some electric furnaces. Magnesia firebrick have been used in Europe prior to 1900, and later for lining reverberatory furnaces for aluminium melting, and it has been reported * that magnesite bricks are favored in Norway for this purpose. No published accounts have appeared. Both bauxite brick and chrome brick have been suggested for use in lining furnaces for aluminium and aluminium alloys, but no actual applications have been made, so far as is known. The main disadvantage of these refractories is their high initial cost. Carbofrax, carbon brick, and zirkite brick have also been used for lining some open-flame furnaces. Hartman [37] has urged the use of carborundum and related products as suitable refractories. Systematic studies of the corrosive action of aluminium and aluminium alloys upon various refractory materials have not been made on a comparative basis. The metallurgical requirements of refractories for use in the aluminium industry have been discussed by the author [47] in another place.

Other Factors Affecting Furnace Operation.—Among other factors, the following are of importance in considering the operation of furnaces, viz., speed of melting; pressure of gases flowing over liquid metal; velocity of furnace gases; volume of gases from various fuels; use of high-pressure gas; relation of weight of charge to melting speed; variation in fuel consumption with size of furnace; burners; and combustion space. These and other factors as applied to brass melting have been discussed in detail by Gillett,[9] and the general principles elucidated by him are applicable to aluminium and aluminium-alloy melting. These factors can be given only very brief treatment here.

With regard to speed of melting, rapid melting and prompt pouring of metal are essential if the oxidation losses are to be kept low. Soaking of metal should be avoided. A method for conserving heat, and consequently for increasing melting speeds, consists in charging the furnace (especially large furnaces) with cold metal after the last heat has been poured at the end of the day, and allowing the metal to absorb part of the furnace heat over night. In general, the best rate of melting is simply the most rapid rate that can be employed. In iron-pot furnace practice, the speed of melting can be increased by the use of a

---
* Private communication, Oct. 7, 1920.

liquid heel, while in reverberatory practice, continuous heats, rather than intermittent, will be found advantageous. The pressure of furnace gases passing over liquid metal is not important in aluminium-alloy work unless zinc-containing alloys are melted. In melting such alloys in open-flame furnaces it is advisable to close all openings in the furnace as much as possible so as to maintain pressure on the bath, and thus reduce zinc volatilization loss, although of course the operating temperature should be lower than the volatilization temperature of zinc.

With regard to the velocity of furnace gases passing over liquid metal, it may be said that the greater the velocity the greater the volatilization and oxidation losses, and at the same time the poorer the fuel efficiency, since there may be insufficient time for the heat to be transferred to the bath. The effect of the volume of gases resulting from combustion is important as related to melting practice, e.g., where a fuel is burned with great excess air supply the melting loss will be high. The employment of high-pressure gas is advocated for the purpose of increasing melting speeds. The size of charge melted should be suitable for the rated capacity of the furnace, and small charges cause low fuel efficiency. In general, there is substantial improvement in fuel efficiency with increase in size of furnaces, and it is good practice to employ a large furnace rather than a number of small ones. The effect of the type of burner used in oil- and gas-fired furnaces upon the fuel efficiency is important, and a burner should be employed that will burn the fuel economically and with a minimum of air. Proper combustion space should be provided, since either too large or too small space is serious from the point of view of melting efficiency. Where the combustion space is too large the furnace is generally too large, and heat is lost by surface radiation from the walls; where the combustion space is too small the fuel will be burned partly outside the furnace, e.g., in the stack, and fuel is lost.

CONSTITUTION OF MELTING CHARGES.

The question of the constitution of the melting charge both in rolling-mill work and in foundry practice is one which is of especial significance in relation to metal and fuel losses, and the exercise of some simple precautions will permit substantial savings to be made. The materials employed in making up heats are necessarily very variable, depending upon local conditions, although most melters endeavor to adhere to a more or less standard charge in which the same relative percentage of materials are used in successive heats. Industrial practice in making up charges of aluminium and certain light aluminium alloys for the production of ingots for rolling or other working and of light aluminium alloys in foundry practice is described below. The preparation of alloys for die castings and permanent-mold castings is also taken up, although less fully, while the matter of scrap and dross recovery by secondary plants is considered in Chapter X.

Melting Charges in Rolling-Mill Practice.—In rolling-mill practice, where substantially pure aluminium is manufactured into sheet, the forms in which aluminium may be charged are variable. In the plants of producers of aluminium, the reduction-cell product is re-melted for the purpose of removing any occluded foreign matter from the furnace bath, and is then cast into rolling ingots. In plants where aluminium is not produced, as well as in the plants of aluminium producers, furnace charges may be exceedingly variable. Thus, the charge may consist of all-primary aluminium pig, all-secondary aluminium pig, all-rolling-mill scrap, all-baled scrap, or combinations of these in various proportions. The charge may vary from 100 per cent primary aluminium pig to 100 per cent scrap of various kinds. Speaking generally, the melting loss is higher when much loose and fine scrap is used than when an all-pig charge is employed, or when heavy scrap is charged. Where light scrap must be re-melted in rolling-mill work, it will be found economical to install a hydraulic or other power baler so that the scrap may be compressed into hard bales. These may be charged to a bath of liquid aluminium and held under the surface, until melted, by a charging tool. Scrap aluminium, when baled, melts quickly, and with practically no melting loss. Loose scrap should never be charged, because it is difficult to immerse and causes heavy oxidation losses. Briquetting has been suggested and tried experimentally for handling coarse and fine " hay " from utensil fabricating operations, and while the resultant dross loss is lower and the briquets made yield a convenient form for charging, the method has not been so far applied in practice.

The conditions governing the large-scale manufacture of aluminium-alloy rolling ingots, e.g., 1.5 per cent manganese-aluminium alloy, are substantially the same as those for aluminium ingots, except that oxidation losses are likely to be higher in the manufacture of alloys owing to the stirring which is necessary to ensure thorough alloying of the additive metal.

Melting Charges in Foundry Practice.—The operating conditions that govern the production of heats of liquid aluminium alloys, and the effects of the constitution of the charge upon dross losses and fuel efficiencies, are substantially the same as given above under Melting Charges in Rolling-Mill Practice. In foundry practice, the object desired is the preparation of a liquid alloy, and superheating it to the correct pouring temperature in the shortest possible time with the lowest melting loss. The principles of alloying and the preparation of light aluminium alloys have been dealt with in Chapter VIII. It may be again pointed out that where metals of very high melting point, e.g., nickel or manganese, are to be alloyed with aluminium, the addition should be made by the use of an intermediate alloy rather than by the substantially pure metal. When the solid cold metals are added to liquid aluminium, as is entirely feasible with zinc and tin, the temperature of the aluminium must be raised very high and the melting period greatly prolonged, or both, in order to effect solution. A metal of low melting point will go rapidly into solution at normal melting temperatures. Some metals, e.g., copper, iron, and nickel, can be readily added in the solid state with entirely satisfactory results, provided that small pieces are used. Large pieces require an excessively long time for solution. The average foundry charge consists of primary aluminium pig, secondary aluminium-alloy pig, foundry scrap, and the necessary additive metal usually in the form of an intermediate alloy. Borings and related fine scraps should not be added to foundry melting charges because they are certain to increase the dross losses. Such scraps should be run into pigs before charging. As a rule, small pieces of scrap should be poked down beneath the surface of a liquid bath, and small scrap should not be charged, as such, into an empty furnace, because too great a surface area is thereby exposed. Oxidation losses are invariably more severe when loose scrap is charged into an empty furnace than when charged to a liquid heel.

The practice of charging foundry-floor sweepings and small pieces of metal from the cleaning and chipping room must be regarded as bad practice; these materials should be run into pigs if they are very small.

The principles governing the make-up of alloys in die-casting and permanent-mold casting practice are substantially the same as in foundry practice, except that generally the alloys are first made up in an alloying furnace, and then cast into pigs. The pigs are then delivered later to the die-casting machine for re-melting. In permanent-mold casting practice, the alloys are melted in furnaces, as in sand work, and the liquid alloys delivered to holding pots near the molds for pouring.

TYPES OF MELTING FURNACES, AND OPERATING DETAILS.

The various types of melting furnaces used in aluminium and aluminium-alloy metallurgy have been described at length by the author [39, 46] in published papers, and these, together with their operating details are taken up briefly in the following pages.

Furnaces in Rolling-Mill Practice.—When speaking of furnaces for melting aluminium, reference is had to melting prior to casting into rolling ingots, although, of course, all direct metal from the reduction cell is re-melted, and either cast into rolling ingots or poured into pigs. Various furnaces have been used for re-melting the reduction-cell metal, but in the United States, open-flame reverberatory furnaces, fired by gas, oil, or coal, are employed principally. A large rectangular Baily electric furnace was used at one time by the Aluminum Co. of America at its Massena, N. Y., plant for this purpose. For melting substantially pure aluminium, prior to casting into ingots for rolling or other working, reverberatory-type furnaces are largely employed in the United States, England, Japan, and Europe. In the United States, the coal-fired reverberatory type appears to have been used mostly, but in later years, gas and oil have been burned in these furnaces. More recently, rotating open-flame gas- and oil-fired furnaces have found favor in some small rolling mills. The employment of the reverberatory-type furnace in rolling-mill work has been found necessary because large capacity is desired, and further, because continuous melting is desirable.

The use of gas-fired reverberatory furnaces in rolling-mill practice has been discussed by the author [25] and M. B. Anderson. The use of the Baily granular-resistor electric furnace for melting aluminium prior to casting into rolling ingots has also been described.[14] For melting aluminium alloys that are to be rolled, the reverberatory furnace is generally preferred, but crucible melting has also been employed.

Furnaces in Foundry Practice.—It is in light alloy foundry practice that the greatest variety of furnaces is employed, and the following types are at present in daily use in aluminium-alloy foundries of the United States: coal-, oil-, and gas-fired reverberatories; oil- and gas-fired stationary and tilting iron-pot furnaces; coal- and coke-fired natural- and forced-draft pit furnaces, using a crucible; coke-, oil-, and gas-fired stationary and tilting crucible furnaces; oil- and gas-fired open-flame tilting and rotating furnaces; and two or three kinds of electric furnaces. The stationary and tilting iron-pot furnaces are favored considerably in this country, particularly for use in large foundries, but open-flame furnaces are being used more widely now than a few years ago. In small foundries, and in foundries where only a minor part of the output is in light aluminium alloys, pit or crucible furnaces are used largely. Electric furnaces for melting light aluminium alloys are now receiving considerable attention, and a few installations have been made. The tendency in recent years toward the use of furnaces of large capacity, such as reverberatories, open-flame barrel-shaped furnaces, and electric furnaces, has come through the enlargement of plant capacity. Where a foundry is turning out from 5,000,000 to 15,000,000 lb. of finished castings per year, melting in units of small capacity requires a large installation, i.e., many furnaces. One type of barrel-shaped open-flame furnace has a melting output of 2,000 lb. of metal per hour, and the advantage in operating larger units than stationary crucibles or iron pots is apparent in foundries where the daily output amounts to thousands of pounds.

Furnaces in Die-Casting Practice.—In aluminium-alloy die-casting work, the alloys are melted in small cast-iron pots, which may be considered as part of the die-casting machine proper. Usually the fuel is gas. In both die-casting and permanent-mold casting work it is standard practice to make up the alloys in an alloying furnace, and run them into small pigs. The prepared alloys are then simply delivered to the die-casting machine for re-melting. Stationary iron-pot furnaces are preferred for alloying, although several other types have been used. In the case of permanent-mold-casting practice, small iron pots fired by gas are preferred by some for melting the alloys prior to casting. In other plants the alloys are melted in cast-iron pots, and then transferred while liquid to graphite-clay holding crucibles.

Furnaces in Secondary Smelting.—In the case of smelting furnaces for aluminium and light aluminium-alloy borings, scrap and dross, practice is not standardized, and many different types of furnaces are employed. For melting borings and other light scrap for the production of so-called " casting aluminium," and in making secondary light aluminium alloys, the following types of furnaces are employed: stationary and tilting iron-pot, reverberatory, pit, and open-flame tilting and rotating furnaces. Strictly speaking, furnaces used for running down borings, dross, and other high aluminiferous scrap should not be called " refining " furnaces, since no refining of the material is accomplished—that is to say, the usual impurities, with the exception of zinc, can not be removed by a melting operation. Experiments have been made by Gillett and James [11] on electric melting of borings, and in the past five years a number of companies have made preliminary studies of the electric melting of aluminium and aluminium-alloy scraps. No large-scale application has yet been made, so far as is known. The primary object in melting aluminium and aluminium-alloy scrap is twofold: (1) simple re-melting for the purpose of pigging clean choice scrap, and (2) obtaining high recoveries on dirty borings and low-grade scrap. The furnace to be employed should be governed by the kind of material melted—at least to a certain extent. Secondary smelting is discussed in Chapter X.

Furnaces for Making Intermediate Alloys.—Intermediate copper-aluminium alloys and other so-called hardeners for foundry practice have been made in various kinds of furnaces. Generally speaking, the stationary iron-pot furnace is preferred for melting the aluminium in the manufacture of say 50 : 50 copper-aluminium alloy, the copper being melted either in pit or in crucible furnaces. Of course, furnaces other than the iron-pot furnace are used for melting the aluminium in the preparation of these alloys. In cases where intermediate alloys are made of aluminium and a metal of high melting point, such as nickel, cobalt, or manganese, an electric furnace may be advantageously employed. A small Rennerfelt electric furnace has been used for this purpose. The Detroit indirect-arc rocking electric furnace has been used for melting aluminium bronze turnings, to which aluminium was subsequently added for the manufacture of 50 : 50 copper-aluminium alloy.

The various furnaces used for melting aluminium and its light alloys are described below.

Pit Furnaces.—Aluminium alloys are still melted in the usual types of brass pit furnaces, particularly in foundries where only a minor part of the output is in these alloys. According to some classifications, under pit furnaces are included all kinds of furnaces, irrespective of the fuel employed, where only a single crucible is used in the furnace, where the melting is done in the crucible, and that crucible is used for pouring the metal. Natural- and forced-draft coke and coal furnaces are generally built in an actual pit below the floor level; so-called pit furnaces, fired by oil or gas, usually are not, although they can be. The latter are generally referred to as crucible furnaces in contradistinction to actual pit furnaces fired by coal or coke, and they are so included in this work. Pit furnaces of various designs, as used in brass practice, have been described at length by Gillett,[9] and their construction need not be considered here.

In American practice, pit furnaces are used in foundry practice for melting aluminium alloys, taking from No. 40 to 400 crucibles, although No. 60 and 80 are the most common sizes. The two latter hold about 60 and 80 lb. of 92 : 8 aluminium-copper alloy, respectively. These furnaces are lined generally with 4 ins. of firebrick. Most pit furnaces are fired by coke, although both bituminous coal and anthracite are used, and natural draft is most frequently employed. From one to eight furnaces are handled per furnace tender, and from 37 to 900 lb. of metal is melted per hour. The fuel consumption is very variable, reported figures varying from 50 to 200 lb. of coal or coke per 100 lb. of alloy, equivalent to furnace efficiencies of from 23 to 8 per cent. In general, the fuel efficiency of the pit furnace is low. Melting costs for fuel in these furnaces may be taken as ranging from 16 to 55 cents per 100 lb. of metal melted, with coke at $6.50 per·ton. As reported to the U. S. Bureau of Mines, from one to six heats are taken from pit furnaces per day, and the melting period varies from 30 mins. to 2 hrs., depending upon the size of the charge. This is equivalent to a melting time of from 1 to 4 hrs. per 100 lb. of alloy. Roughly, a pit furnace, fired by coke, should give at least six heats of 100 lb. per 9-hr. day, with an average melting period of 75 minutes.

The life of linings in pit furnaces varies from 4 to 24 months, equivalent to 400 to 1,800 heats, with an indicated average of 1,000 heats. The life of crucibles in pit furnaces varies from 28 to 100 heats, with an indicated average of 42 heats, equivalent to about 2,500 lb. of metal. The gross melting loss in pit furnaces varies from 2 to 3 per cent, with an indicated average of 2.3 per cent. The net melting loss, assuming recovery from dross and skims, may be taken as about 1.4 per cent. On the whole, pit furnaces may be regarded as fairly satisfactory for melting aluminium alloys from the point of view of melting losses, but they are slow in operation, troublesome to handle, and exceedingly inefficient as to fuel.

Crucible Furnaces.—Stationary crucible furnaces are used considerably in foundries for melting aluminium alloys, but the tilting type is employed only slightly. Both types are fired by gas or oil. In the stationary crucible furnace, a graphite-clay crucible is set in a furnace made by lining an iron shell with refractory material. The shell may be set down partly in a pit, or it may be above the level of the floor, generally the latter, and built as a separate unit. The tilting crucible furnace consists of an iron shell lined with refractory material, in which a graphite-clay crucible with a molded pouring lip is placed; the shell is set on trunnions and fitted with the necessary tilting mechanism. The capacity of the ordinary stationary crucible furnace is small, these being built up to No. 600 crucibles, but No. 60 to 100, with capacities of 60 to 100 lb. of 92 : 8 aluminium-copper alloy are common sizes. In foundry practice, the tilting furnace taking a No. 80 crucible is the usual size. Crucible furnaces are ordinarily run on intermittent heats, and the practice of employing a liquid heel may be followed. In stationary crucible furnaces there is ample opportunity for the products of combustion to come into contact with the alloy during melting, and contamination from the fuel may occur, which would not be possible in iron-pot melting. In the ordinary stationary crucible furnace the crucible is heated over its exterior surface by the combustion of the fuel from burners, and the waste gases are discharged through a large hole in the furnace cover. The flames are projected tangentially, so that they are given a rotary and upward motion around the crucible, and the incandescent gases are in intimate contact with the alloy during melting. In tilting crucible furnaces the products of combustion may or may not have opportunity to come into contact with the alloy during melting, depending upon the design of the furnace.

In American practice, stationary crucible furnaces are generally small units, and such furnaces are installed usually in foundries of small output as single units. They are rarely run in a battery. These furnaces generally have a 4-in. lining inside the furnace shell. Cast-iron, firebrick, and molded-clay covers are used on crucible furnaces, the last two having generally a hole in the center for the escape of the combustion products. Both oil and gas are used for fuels, oil being the more common. Details of the burners used and the air supply can not be gone into here, but low-pressure air is used on most of the furnaces. In actual installations, from one to three furnaces are handled per furnace tender, and from 110 to 200 lb. of metal are melted per hour. The fuel consumption in stationary crucible furnaces is variable, reported figures varying from 0.54 to 15 gals. of fuel oil per 100 lb. of alloy, equivalent to furnace efficiencies of from 2.3 to 64.8 per cent. On the whole, the fuel efficiency in stationary crucible furnaces is low, and the exceptionally high efficiency reported is open to question. As a rule, on the basis of reported figures, the fuel consumption in these furnaces will vary from 10 to 15 gals. of oil per 100 lb. of 92 : 8 aluminium-copper alloy, with an average melting cost for fuel of $1.

From 5 to 12 heats per 9-hr. day are taken from stationary crucible furnaces in practice, and the melting period varies from 30 mins. to 1 hr., depending upon the size of the charge. This is equivalent to a melting time of from 45 mins. to 3 hrs. per 100 lb. of charge. Roughly, a stationary crucible furnace, fired by oil, using a No. 100 crucible taking about 100 lb. of alloy, should give at least 7 heats per 9-hr. day, with an average melting period of about 1 hr. The life of linings in stationary crucible furnaces varies from 3 months to 2 years, equivalent to 150 to 9,000 heats, with an indicated average of 3,000 heats. The life of crucibles in stationary crucible furnaces varies from 18 to 100 heats, as reported, with an indicated average of 49 heats, equivalent to about 2,900 lb. of metal melted. The gross melting loss in these furnaces varies from 1 to 6 per cent, as reported, with an indicated average of 3.6 per cent. The net melting loss may be placed at 2 per cent. On the whole, crucible furnaces, both stationary and tilting, may be regarded as fairly suitable for melting light aluminium alloys in foundry practice for some purposes, but they are inefficient as to fuel, and the melting losses are rather high. Scant data are available as to the performance of tilting crucible furnaces melting aluminium alloys under foundry conditions, but in general the data given above as to stationary crucible furnaces apply.

Stationary Iron-Pot Furnaces.—As mentioned previously, stationary iron-pot furnaces are favored considerably in foundries for melting light aluminium alloys. These furnaces may be run as separate units, or in a battery of several pots to a bank; the practice of employing two or three pots to a bank has gained favor. Oil or gas are generally used as fuel. At the present time, stationary iron-pot furnaces are run without covers in most foundries, although some operators are still strongly disposed to use solid cast-iron covers on small units. Stationary iron-pot furnaces are built in various capacities up to 400 lb., and 250 lb. is a common size. The pots may either be round or rectangular, but are mostly of the former shape. In the stationary iron-pot furnace, the alloy to be melted is placed in the pot, which is heated over its exterior surface by the combustion of the fuel in the space between the furnace walls and the pot. Normally, the products of combustion do not have an opportunity to come into contact with the alloy during melting. Ordinarily, a liquid heel of 50 to 150 lb. is left in the pot to conserve the heat and also facilitate alloying, and additional cold metal is charged in from time to time during the day. When a charge is melted, and at the proper temperature for pouring, the alloy is ladled out of the pot and poured.

In the case of single units, the furnace consists of a round, square, or rectangular shell lined with refractory brick, while the top is covered over with a refractory roof, leaving a hole for setting the pot. The latter is hung from the top of the furnace, and may be supported by wedge-shaped bricks or by a refractory stool below. One burner is used in single units, and it should be so placed that the flames are not directed immediately against the pot but rather around it, and the products of combustion are discharged through a small flue at the rear. In some furnace installations the products of combustion are discharged through the flue directly into the air, while in others the small flue leads to a large connecting flue as in pit furnaces, and thence to a stack. With one furnace, a small stack may be used. Where two or more pots are placed in one furnace a burner is used at each end. Charging of the pots is so arranged that the burners are used alternately. Thus, in the case of a 2-pot furnace, with a charge of liquid metal in one pot and of cold metal in the other, the burner nearest the first pot would be operating, while the second pot would be pre-heated by the hot gases. After pouring the first pot it would be charged with cold metal, the first burner turned off and the second one started. This principle is also applied in some double-chamber tilting crucible furnaces and in double-chamber open-flame tilting furnaces, such as the Rockwell. At the present time the iron-pot furnace is the one most extensively employed from the point of view of the quantity of metal melted. Fig. 101 shows the top view and cross section of a single-unit stationary iron-pot furnace, and Fig. 102 is a view of an installation of stationary iron-pot furnaces in a foundry.

In the United States, the capacity of stationary iron-pot furnaces varies from 100 to 400 lb. of 92 : 8 aluminium-copper alloy in the case of circular pots, and from 600 to 1,000 lb. for the rectangular pots. The circular pot holding 300 lb. of alloy is a common size; this may be bowl-shaped, 18 ins. in diameter by 18 ins. deep. A typical rectangular pot is one 16 ins. wide by 37 ins. long by 26 ins. deep; this holds from 1,000 to 1,200 lb. of alloy. Iron-pot furnaces of the stationary type are usually lined with 4 ins. of firebrick, although silica brick, carborundum fire sand, and high-temperature cements are used in practice. Covers are generally used on single-unit iron-pot furnaces; these are simply dome-shaped covers made of cast iron about 1 in. thick. In large installations of these furnaces, covers are not employed as a rule. Oil is the preferred fuel for use in iron-pot furnaces; natural gas is used to a less extent, while illumi-

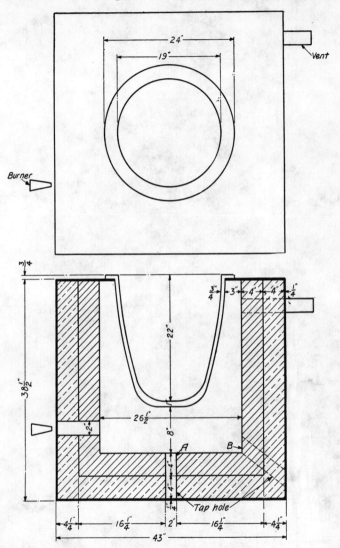

FIG. 101.—*Top view and cross-section of a single-unit stationary iron-pot furnace*

nating (city) gas and blue water gas are employed in a few installations. Details of the burner equipment and air supply

FIG. 102.—View showing installation of stationary iron-pot furnaces.

for iron-pot furnaces of the stationary type can not be included here, but a typical installation may be cited: pressure of air at the burner, 1.5 to 2 lb.; pressure of oil at the burner, 25 to 30 lb.; Anthony burner.

From one to eight furnaces are handled per furnace tender in practice, and from 100 to 2,400 lb. of alloy per hr. are melted. The fuel consumption in these furnaces is very variable, reported figures varying from 1.25 to 5 gals. of oil per 100 lb. of alloy, and from 150 to 1,000 cu. ft. of natural gas, equivalent to furnace efficiencies of from 5.7 to 57.3 per cent. In general, the average fuel efficiency in these furnaces is higher than in pit or crucible furnaces. Melting costs for fuel in iron-pot furnaces of the stationary type, using oil, vary from 10 to 40 cents per 100 lb. of alloy. From 3 to 10 heats are taken per day, and the melting period varies from 45 mins. to 3 hrs., depending upon the size of the charge. This is equivalent to a melting time of from 12 mins. to 1 hr. per 100 lb. of alloy. Roughly, it may be said that at least seven 300-lb. heats should be obtained per furnace per 9-hr. day, with an average melting period of 1.25 hrs. A liquid heel is used in iron-pot furnace practice where it is generally recognized as advantageous. In the majority of plants, a heel of 50 to 100 lb. will be retained in the furnace after drawing a heat, and cold metal is charged to the heel. The life of linings in stationary iron-pot furnaces is variable, from 3 to 12 months being reported. This is equivalent to from 225 to 1,300 heats, with an indicated average of 6 months and about 1,000 heats. The life of iron pots varies from 24 to 80 heats, with an indicated average of 47 heats, equivalent to about 14,000 lb. of metal for a 300-lb. pot. The gross melting loss in these furnaces varies from 2 to 6 per cent, with an indicated average of 4.2 per cent. This is equivalent to a net melting loss of 2.5 per cent. Iron-pot furnaces may be regarded as fairly satisfactory for melting aluminium alloys in foundry practice, particularly where the output is fairly large. They are easy to handle and the labor cost for melting should be low. In actual practice they are seen to be more efficient as to fuel consumption than pit or crucible furnaces, although the melting losses appear to be higher.

Tilting Iron-Pot Furnaces.—Tilting iron-pot furnaces are used considerably in foundry practice for melting aluminium alloys. They are ordinarily fired by oil or gas and invariably operated as single units; their general construction is similar to that of the stationary iron-pot furnace, viz., a cast-iron pot is mounted in a furnace shell and heated over its exterior surface by combustion of the fuel. As a rule, the capacity of the tilting furnace is larger than the size in which the stationary furnace is built. Thus, furnaces with capacities up to 600 lb. of 92 : 8 aluminium-copper alloy are built, and 300 lb. is a common size.

In contradistinction to stationary iron-pot furnaces, tilting iron-pots are mostly run on intermittent heats, i.e., there is no liquid heel utilized, and the inside of the pot is scraped thoroughly after each heat. It is easier to clean out a tilting pot than a stationary one. On the other hand, the practice of employing a liquid heel is also in force at some plants having the tilting furnace. The pots used in tilting furnaces are always round in shape. The products of combustion do not have much opportunity to come into contact with the alloy during melting, since the pot is heated over its exterior surface by combustion of the fuel; but where the waste furnace gases are discharged through openings at the top of the furnace adjoining the upper rim of the pot there is opportunity for the gases to come into contact with the metal. However, in both stationary and tilting iron-pot furnaces the metal may be regarded, for all practical purposes, as effectively protected from the products of combustion, in contradistinction to the open-flame furnaces, where the incandescent combustible gases are directly in contact with the metal during the melting period. The tilting furnace presents some definite advantages over the stationary type, since in the case of the latter the liquid alloy must be ladled out into pouring vessels, with resultant additional oxidation and also much spattering and spilling. Under foundry conditions, the quantity of furnace scrap (spills and pouring skims) is usually less when the tilting furnace is used. The following information supplied by a furnace-maker applies to both stationary and tilting iron-pot furnaces in general:

The furnace (a commercial design) is of the tilting type using a cast-iron melting pot, built in two sizes of 100- and 300-lb. capacity. The linings are made of machine-pressed high refractory firebrick, and a cast-iron hinged cover is used. For the pot of 100-lb. capacity, the outside diameter of the flange is $17\frac{13}{16}$ ins., and the outside diameter under the flange is $15\frac{13}{16}$ ins.

The overall depth of the pot is $14\frac{3}{4}$ ins. For the 300-lb. pot, the outside diameter of the flange is $23\frac{1}{8}$ ins., and the outside diameter under the flange is $20\frac{1}{8}$ ins. The overall depth is $15\frac{5}{8}$ ins. The weight of the 100-lb. pot is 100 lb. and that of the 300-lb. pot is 225 lb. Natural or artificial gas, fuel oil, or kerosene may be burned. The furnace of 100-lb. capacity consumes about 4 gals. of fuel oil per hour, 4.5 gals. of kerosene, 600 cu. ft. of natural gas, or 850 cu. ft. of artificial gas per hour. The fuel consumption on the 300-lb. furnace is approximately 60 per cent more per hour than on the 100-lb. furnace. When burning oil, the oil pressure is 40 lb. per sq. in., and the air pressure about 4 oz. When burning gas, the air pressure is 4 oz. and the gas pressure 4 oz. In the case of the 100-lb. furnace, the time required for the first heat, when the furnace is started from the cold, is 75 mins.; additional heats, when the furnace is run continuously, require 60 mins. per heat. For the 300-lb. furnace, the time required for the first heat, when the furnace is started from the cold, is 105 mins.; additional heats, when the furnace is run continuously, require 75 mins. The furnace consumes about 4.5 gals. of oil per 100 lb. of metal melted. Seven heats can be obtained in an 8-hr. working day, and a liquid heel is usually recommended. When the furnace is run full time every working day, the lining will require replacement once every 6 months. It is stated that 50 to 55 heats can be obtained from the cast-iron pots. The average thickness of side walls is $\frac{3}{4}$ in., and the bottom $\frac{7}{8}$ to 1 in. The failure of iron pots is attributed to stretching and scaling. Three furnaces require one furnace tender. Fig. 103 shows an installation of tilting iron-pot furnaces in a large foundry. In general, the performance of tilting iron-pot furnaces is similar to that of the stationary type, as to operating details, and further data need not be included here.

Open-Flame Furnaces Other than Reverberatories.—Open-flame furnaces of various designs are favored by some foundrymen for melting light aluminium alloys, chiefly on account of the rapidity of melting. These furnaces, also called direct-flame furnaces, consist of a sheet iron or steel shell, either egg-shaped, pear-shaped, or cylindrical, lined with a refractory material and mounted on trunnions. The principal types include the following: egg-shaped tilting furnaces; pear-shaped (upright) tilting furnaces; cylindrical stationary furnaces; cylindrical

FIG. 103.—Installation of tilting iron-pot furnaces in a large foundry.

tilting furnaces; and cylindrical rotating and tilting furnaces. They are fired by oil or gas, and the combustion of the fuel takes place in the space directly above the surface of the metal; the metal is in intimate contact with the products of combustion. In the case of the single-unit small pear-shaped (upright) furnace, two burners are placed in the side at a point midway between the charging door and the pouring spout. The furnace is charged through the door at the top, and the metal is poured through the spout on tilting the furnace to the proper position. The pear-shaped shell is mounted on trunnions. The open-flame egg-shaped furnaces may be lined with firebrick or with a rammed lining of ganister and fireclay. Usually, rather small furnaces are employed for melting aluminium alloys, a furnace 42 ins. in diameter being a common size, although those up to 75 ins. are used. For the 42-in. Hawley (Schwartz) furnaces, 300 cu. ft. of air per min. is consumed under a pressure of 12 oz. (with two $1\frac{5}{8}$ in. diameter tuyères). Oil is used under a pressure of 25 to 30 lb., and is pre-heated to 50° C. to ensure an easy flow. When gas is used as fuel the pressure should be about 0.25 lb. more than the air pressure. The capacity of the common open-flame pear-shaped tilting furnace for aluminium alloys varies from 100 to 2,000 pounds.

Another type of open-flame furnace is the barrel-shaped tilting furnace. Its construction is similar to that of the pear-shaped furnace, but it is simply different in shape. A charging door is built in the top and a pouring spout on one side. One burner may enter through one trunnion, or one or two burners may be placed through the side walls near the top. A barrel-shaped rotating and tilting furnace has recently been devised. It is made to rotate, but it tilts at right angles to the direction of rotation. One burner is placed so as to project the flame through a hole at one end of the furnace and out through the pouring spout at the other end. This furnace is now used for melting aluminium for pouring into rolling ingots, for melting aluminium alloys in foundry practice, and for running down borings, drosses, and similar scraps. Some other types of open-flame furnaces are on the market, but their performance for melting light aluminium alloys is not known. Fig. 104 is a sketch showing end views of the Schwartz cylindrical stationary open-flame furnace, and Fig. 105 is a photograph of the open-flame rotating and tilting furnace (U. S. Smelting type).

Some data are available as to the operating performance of open-flame furnaces, and most of those given below are for pear-shaped (upright) open-flame tilting furnaces, and the remarks made should not be applied indiscriminately to all types of open-flame furnaces. Open-flame furnaces are very

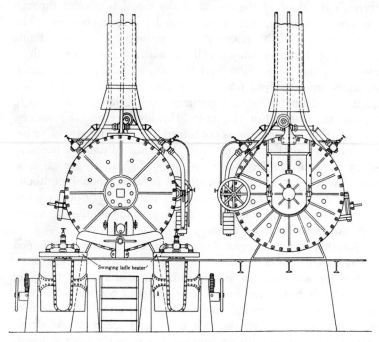

FIG. 104.—*End views of the Schwartz cylindrical stationary open-flame furnace* (*Edward H. Schwartz*).

variable as to capacity, but, as indicated above, the pear-shaped (upright) furnace having a diameter of 42 ins. and taking from 200 to 300 lb. of 92 : 8 aluminium-copper alloy appears to be the most favored. These furnaces may be lined with 4 ins. of firebrick, although carborundum fire sand is used considerably. In one installation, a rammed lining of ganister and fireclay, 3 ins. thick, is employed. Natural gas and oil are used for the fuel, the latter being preferred. In this installation the pressure of the air at the burner is from 16 to 18 ozs., and the pressure of the oil is 40 lb., using a burner supplied with the furnace. In practice, one or two furnaces are handled per furnace tender, and from 200 to 600 lb. of 92 : 8 aluminium-copper alloy are melted per hour.

The fuel consumption in these furnaces is variable; in one furnace it was 450 cu. ft. of natural gas per 100 lb. melted, while in another it was 2 gals. of oil. These figures are equivalent to efficiencies of 12.7 and 17.5 per cent, respectively. In general, the fuel efficiency in open-flame furnaces of most types appears to be substantially higher than in the furnaces heretofore discussed. Melting costs, for the two furnaces just mentioned, are 18 and 16 cents per 100 lb. of alloy. Open-flame furnaces

FIG. 105.—*Open-flame rotating and tilting furnace; U. S. Smelting type.*

are the most rapid fuel-fired melting furnaces. Thus, in one plant, with an oil-fired furnace of the pear-shaped (upright) tilting type, six 300-lb. heats are obtained in 9 hrs., with an average melting period of 30 mins. This is equivalent to a melting time of 10 mins. per 100 lb. of alloy. The life of linings in these furnaces varies, of course, depending upon a number of factors. In one plant, a 4-in. firebrick lining had a life of about 3 months, giving 600 heats, melting 200 lb. of 92 : 8 aluminium-copper alloy per hour, 9 heats per day. This is equivalent to about 125,000 lb. of metal melted. In another plant, a 300-lb. furnace with a 3-in. lining of rammed ganister and fireclay gave 6 heats per day, and lasted four years. This is equivalent to 7,200 heats and a melting output of about 2,100,000 lb. of metal.

426 Metallurgy of Aluminium and Aluminium Alloys

The average gross melting loss in open-flame furnaces is generally high, although this is governed principally by the furnace atmosphere and the temperature. This ranges from 4 to 8 per cent in practice, with an indicated average of 6.7 per cent. This is equivalent to a net melting loss of 4 per cent. Open-flame furnaces are very rapid melting furnaces, and their efficiency as to fuel consumption is relatively high. They give rise to heavy oxidation losses unless the furnace atmosphere is lean in oxygen, and it may be that such control of the atmosphere will result in heavy fuel losses.

Reverberatory Furnaces.—Reverberatories are used to some extent for melting aluminium alloys in foundries, but chiefly for melting aluminium for casting into rolling ingots. Their capacity is small when compared with reverberatory furnaces used in brass and copper practice, and they are generally built to hold from 500 to 6,000 lb. of aluminium. There are two general types of reverberatory furnaces, viz., (1) those in which the liquid metal is ladled out through a door, and (2) those in which a runner is used for tapping. Furnaces of large capacity are required in aluminium rolling-mill practice, and it is probably for this reason and the fact that they can be readily built by workmen around a plant, that they have been used for ingot casting. In foundry practice, the reverberatory furnace is applicable where the output is large and continuous, but it is necessary that the production of castings be so planned that large heats of metal are not held in the furnace for long periods of time. The "soaking" of aluminium heats in rolling-mill practice does not appear to be so deleterious to the quality of the metal as it is to aluminium alloys in foundry practice. The reverberatory furnace is also of value in foundry practice because it permits the charging of large scrap, such as defective crank-cases and the like, without breaking. In smaller furnaces, such as the iron-pot and stationary crucible furnaces of the usual sizes, it is necessary to break up all large scrap to enable charging it. Reverberatories are also used for running down borings and other scrap in the production of secondary aluminium and aluminium-alloy pig.

The common type of reverberatory furnace used for melting aluminium and aluminium alloys consists of a firebrick shell bound together with steel girders (channels and I-beams).

Aluminium-Alloy Melting Practice

The hearth is hollow, and is generally built of firebrick, while the roof is arched over from one side to the other, and often arched down from front to back in order to deflect the flame and the hot gases from combustion down upon the metal. Reverberatories are fired with coal, oil, or gas. When coal is used, the fuel is burned upon a grate placed in a firebox at the front of the furnace; natural draft is generally used. When fired by oil or gas, the burners are usually placed in the sides of the furnace. The products of combustion are led to a stack by way of a flue at the rear. From two to four burners are employed. Fig. 106 shows a reverberatory furnace. Reliable data as to the

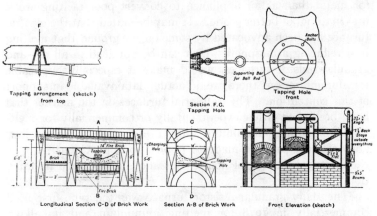

Fig. 106.—*Reverberatory furnace (Rhodin).*

performance of reverberatory furnaces on aluminium alloys are scarce, but in general the remarks made above as to pear-shaped open-flame tilting furnaces will apply here.

Electric Furnaces.—Little information is available as to the performance of electric furnaces in melting aluminium and its alloys, and at the present time there is as little known about the electric melting of aluminium and aluminium alloys as there was about electric brass melting ten years ago. Gillett [30, 43] has discussed the present status of the electric melting of aluminium, and has described the furnaces now in operation. He considers that since it is only in the zinc-containing aluminium alloys that there is any considerable volatilization loss, there is little opportunity for the electric furnace in the aluminium field from the point of view of volatilization, because only a relatively small quantity of aluminium-zinc alloys are now made. Provided that the tendency toward the employment of ternary aluminium-copper-zinc alloys continues, the electric furnace will be more important in the aluminium-alloy foundry on the basis of prevention of volatilization. At the same time there are certain advantages in cleanliness and speed of melting obtainable in the electric furnace, and many foundrymen are now regarding it with favor. The idea has been held by some that a superior quality of metal would result from electric-furnace melting, but the electric furnace is to be regarded rather as a melting medium for metal than as an appliance to convert poor melting stock into material of better grade. It may be stated that the electric furnace has been advocated by some on the ground that melting in a reducing atmosphere which can be obtained in all arc and granular-resistor furnaces, would make it especially applicable for aluminium, but there are no actual data available in support of this contention. The industrial furnaces on the market that have been used either experimentally or commercially for melting aluminium or aluminium alloys are discussed briefly below, and reference may be made to publications by Gillett [30, 43] on the electrical melting of non-ferrous alloys for detailed information.

The Baily granular-resistor furnace was the type first used commercially in 1918 for melting aluminium and its alloys. The 500-kw. furnace at the Massena, N. Y., plant of the Aluminum Co. of America was the first installation, although the tilting type of this furnace had been tried experimentally for alloys before this time. This furnace was installed to replace an oil-burning reverberatory furnace for re-melting the reduction-cell product for casting aluminium into pigs, but it is reported that the furnace has been discarded for this purpose. The large rectangular Baily furnace consists, briefly, of an iron shell with a lining of high-grade firebrick, backed up by insulating material between the brick and the shell. The hearth is bowl-shaped, and has a capacity of 3 to 4 tons of liquid aluminium, the daily output for 24-hr. operation being 20 tons. There are doors at both ends of the furnace, together with three rabbling doors on the side opposite the tap-hole. This furnace is stationary, and the metal is tapped by pulling out a plug from the tap-hole.

Running lengthwise of the furnace are two resistor troughs made of refractory material and filled with resistor carbon or graphite. The troughs are supported on brick piers. Electric current is sent through the resistor troughs, contact being made through copper terminals at the ends, and the heat generated in the trough is radiated largely to the furnace roof and then reflected down upon the hearth. Fig. 107 is a view of this furnace which has been described by Miller.[14]

FIG. 107.—*Baily rectangular hearth electric furnace* (*Baily*).

The Baily tilting electric furnace is employed in a few foundries for melting aluminium alloys. The standard 105-kw. furnace consists of a cylindrical steel shell mounted on trunnions, and fitted with mechanism so that it may be tilted by hand. A circular refractory trough filled with crushed carbon-resistor material, and mounted on a series of refractory piers, is the heating element. Current connection is made to the resistor element by means of copper leads terminating in graphite blocks packed against the resistor material. The metal is melted on a hearth below the trough by the heat reflected from the incandescent resistor. The furnace shell is about 7 ft. high by 6 ft. diameter, and the furnace holds 500 lb. of 92 : 8 aluminium-copper alloy. The hearth is made of carborundum fire sand, and the roof and side walls are lined with corundite brick. A backing of infusorial earth is packed between the lining and furnace shell. The current consumption is given as about 600 kw.-hrs. per ton of aluminium alloy melted. The furnaces are built in several sizes and have been described by Baily.[12, 19] One of these furnaces is in regular operation in the foundry of the Dayton Engineering Laboratories, Dayton, Ohio, for melting aluminium alloys. A test run on this furnace gave an output

FIG. 108.—*Baily tilting type electric furnace* (*Baily*).

of about 1.5 tons in 8 hrs. of actual melting, plus one hour for pre-heating, or 7 to 8 heats of 425 lb. each. The average power consumption is stated to be 625 to 675 kw. per ton, and the average power input was only 80 kw. A small 50-kw. Baily furnace was in operation at the plant of Landers, Frary and Clark, New Britain, Conn. This furnace gave an output of 0.5 ton of metal per day, melting 9 hrs. and being pre-heated 4 hrs. The net metal loss was about 0.7 per cent. Fig. 108 shows a view of this furnace.

The Detroit indirect-arc rocking electric furnace has been used for melting aluminium alloys in foundry practice. This furnace consists essentially of a cylindrical steel shell lined with a layer of corundite brick, which is backed by a course of less refractory brick chosen for its heat-insulating properties, and this, in turn, is backed by brick of infusorial earth next to the iron shell. This furnace is built in various sizes, but the 1-ton size for brass takes about 700 lb. of 92 : 8 aluminium-copper alloy. The furnace proper is mounted on rollers and ring gears, which permit it to be rocked through any desired arc of revolution up to 200 degrees. The rocking motion is actuated by a small induction motor through reducing gears, and the action of the motor is controlled by an automatic reversing switch which may be set to give the desired movement. The heat is generated by an electric arc from two horizontal graphite electrodes placed axially in the center of the furnace, and meeting in the center of the chamber. The electrodes are controlled by hand wheels, but automatic regulation may be applied. During charging of metal, the electrodes are run back out of the way of the charging door. To operate, the furnace is closed, the electrodes are brought to the operating position, and the arc is started. The metal is melted by conduction from the refractory lining, as well as by direct radiation from the arc. Rocking is started after the metal commences to soften, first through a small angle, and finally reaching the maximum angle as the metal enters the stage of superheating. The movement is two complete oscillations per minute. The charging door and spout are at the center on one side of the furnace, and the furnace may be charged mechanically from above by rotating the furnace so that the door is on top. The indirect-arc rocking furnace (Detroit type) has been described in detail by Gillett and Rhoads [13] and by St. John.[20] A standard Detroit rocking indirect-arc furnace at the C. B. Bohn Foundry Co., Detroit, Mich., has been used for melting aluminium alloys; the net melting loss is given as 0.85 per cent. This furnace has also been used for making 50 : 50 copper-aluminium alloy.

The G. E. type electric furnace is used by the General Electric Co. at several of its plants for melting aluminium and its alloys. This furnace consists of a shallow hearth in the middle of a rectangular furnace shell. Two heating troughs placed on each side of the hearth, and separating the hearth from the side walls, are the source of heat. The heating troughs are filled with granular coke, and two long cross electrodes are placed in the bottom of the troughs. Two vertical electrodes make contact with the cross electrodes through wearing blocks, and the ends of the vertical electrodes are covered with the granular resistor material. The furnace is inherently two-phase, each pair of vertical electrodes being connected to one phase. Heat is developed in the resistor troughs by passage of the current and by contact arcs at the bottom of the vertical electrodes. The heat is radiated to the arched roof and thence reflected to the hearth, and the body of granular material conducts heat to the sides and bottom of the hearth. This furnace has been described in detail by Collins.[24] The General Electric Co. uses a small 75-kw. furnace like its older type of electric brass furnace, with a capacity of 75 to 100 lb., for melting aluminium for rotor castings. This company also has a larger furnace of the same type (G. E. smothered-arc furnace) of about 500-lb. capacity, rated at 150 kw., for melting light aluminium alloys in foundry practice. A much larger furnace of the same type, to hold several tons of metal, has been designed and is to be tried experimentally. The General Electric furnaces for use on aluminium are the same as those for brass, except that in the former the transformer capacity and power input are lower for the same size shell.

A number of other electric furnaces have been used experimentally or commercially for melting aluminium alloys, of which the Rennerfelt stationary indirect-arc type has been employed for the preparation of so-called "hardeners." The principle of heating is by reflected heat from an electric arc, and the furnace has been described by Vom Bauer [10] and others. The Rennerfelt electric reverberatory furnace [28] has been suggested for aluminium melting. This furnace has two melting hearths, and the heat is generated by a resistor trough, which is heated by an electric arc from vertical electrodes. The arc may be smothered in the resistor, as in the General Electric furnace, or open. The vertical ring induction furnace (Ajax–Wyatt) has not been used for aluminium-alloy melting, although it is understood that it is to be tried experimentally for this purpose. The direct-arc type electric furnace, of which the Snyder furnace is the only one, of many of this type, applied to copper alloys, has not been used for aluminium alloys.

Aluminium-Alloy Melting Practice

SELECTED BIBLIOGRAPHY.

The following references include certain papers dealing with the general aspects of metal-melting practice as well as papers concerned principally with aluminium and aluminium-alloy melting.

1. Krom, L. J., The development of melting furnaces, *The Metal Ind.*, vol. 7, 1909, pp. 287–289; 324–326; 358–361; 404–406; 436–440; and vol. 8, 1910, pp. 80–81.
2. Desch, C. H., Some common defects occurring in alloys, *Jour. Inst. of Metals*, vol. 4, 1910, pp. 235–246.
3. Sieverts, A., and Krumbhaar, W., Über die Löslichkeit von Gasen in Metallen und Legierungen, *Ber. deut. Chem. Gesell.*, vol. 43, 1910, pp. 839–900.
4. Sieverts, A., Über Lösungen von Gasen in Metallen, *Zeit. für Elektrochem,*. vol. 16, 1910, pp. 707–712.
5. Barnes, E. A., Non-ferrous foundry economics and refinements, *Trans. Amer. Inst. of Metals*, vol. 5, 1911, pp. 90–113.
6. Woods, C. F., Report of official chemists of the American Institute of Metals, *Trans. Amer. Inst. of Metals*, vol. 6, 1912, pp. 1–11; discussion, pp. 11–43.
7. Guichard, M., and Jourdain, P.-R., Sur le gaz de l'aluminium, *Compt. Rend.*, vol. 155, 1912, pp. 160–163.
8. Collins, J. W., Suggestions for making aluminium castings, *The Foundry*, vol. 42, 1914, pp. 67–68; abst. of paper before the Detroit Foundrymen's Assoc.
9. Gillett, H. W., Brass-furnace practice in the United States, U. S. Bureau of Mines Bull. 73, 2nd ed., June, 1916.
10. Vom Bauer, C. H., The Rennerfelt electric arc furnace, *Trans. Amer. Electrochem. Soc.*, vol. 29, 1916, pp. 497–504.
11. Gillett, H. W., and James, G. M., Melting aluminium chips, U. S. Bureau of Mines Bull. 108, August, 1916, 88 pp.
12. Baily, T. F., Resistance type furnace for melting brass, *Trans. Amer. Electrochem. Soc.*, vol. 32, 1917, pp. 155–163.
13. Gillett, H. W., and Rhoads, A. E., Melting brass in a rocking electric furnace, U. S. Bureau of Mines Bull. 171, July, 1918; and, A rocking electric brass furnace, *Chem. and Met. Eng.*, vol. 18, 1918, pp. 583–590.
14. Miller, D. D., The remelting of aluminium pig in the electric furnace, *Chem. and Met. Eng.*, vol. 19, 1918, pp. 251–254.
15. Anderson, R. J., The practice of melting and casting aluminium, *The Foundry*, vol. 46, 1918, pp. 104–106; 164–166.
16. Hill, E. C., Aluminium: its use in the motor industry in England, *The Metal Ind.*, vol. 16, 1918, pp. 543–546; and vol. 17, 1919, pp. 126–127.
17. Stull, R. T., Behavior under brass foundry practice of crucibles containing Ceylon, Canadian and Alabama graphites, *Jour. Amer. Ceramic Soc.*, vol. 2, 1919, pp. 208–226.
18. Anderson, R. J., Blowholes, porosity, and unsoundness in aluminium-alloy castings, U. S. Bureau of Mines Tech. Paper 241, December, 1919, pp. 11–13.
19. Baily, T. F., Electric furnaces of the resistance type for melting non-ferrous alloys, *Trans. Amer. Inst. of Chem. Engrs.*, vol. 12, part I, 1919, pp. 29–47.
20. St. John, H. M., The Detroit rocking electric furnace for melting brass and bronze, *Trans. Amer. Inst. of Chem. Engrs.*, vol. 12, part I, 1919, pp. 81–94.

21. Symposium, Occlusion of gases by metals, Trans. Faraday Soc., vol. 14, 1919, pp. 173–239.
22. Lea, F. C., The founding of aluminium, *The Metal Ind.* (London), vol. 15, 1919, pp. 309–311; abst. of paper before the Royal Aeronautical Soc., April, 1919.
23. Bregman, A., Metallurgical furnaces, *The Metal Ind.*, vol. 17, 1919, pp. 159–162; 218–220.
24. Collins, E. F., Melting of some non-ferrous metals and their alloys in the electric furnace, *Jour. Cleveland Eng. Soc.*, vol. 11, 1919, pp. 293–320.
25. Anderson, R. J., and Anderson, M. B., Aluminium rollimg-mill practice, *Chem. and Met. Eng.*, vol. 22, 1920, pp. 489–491; 545–550; 599–604; 647–650; 697–702.
26. Prentiss, F. L., Modern foundry for aluminium castings, *The Iron Age*, vol. 105, 1920, pp. 535–539.
27. Hill, C. W., Thomas, T. B., and Vietz, W. B., Investigation of brass foundry flux, Trans. Amer. Inst. of Min. and Met. Engrs., vol. 64, 1920, pp. 662–673.
28. de Fries, H. A., Rennerfelt electric reverberatory furnace, *Chem. and Met. Eng.*, vol. 22, 1920, pp. 280–281.
29. Eastick, T. H. A., Melting furnaces and equipment for non-ferrous metals, *The Metal Ind.* (London), vol. 17, 1920, pp. 149–151.
30. Gillett, H. W., Electrical melting of alloys, *The Foundry*, vol. 48, 1920, pp. 177–180; 229–231, 238; 275–281; 319–323; 362–366; 400–406: 445–453, 459; 486–491; 531–539; 571–574; 612–614; 656–658; and 693–695.
31. Cone, E. F., Electric melting in non-ferrous industry, *The Iron Age*, vol. 105, 1920, pp. 655–656.
32. Anderson, R. J., Analysis of losses in aluminium shops, *The Foundry*, vol. 48, 1920, pp. 989–992; and *idem*, vol. 49, 1921, pp. 16–21; 54–57; 104–111; 143–147; 188–191; and 235–239.
33. Anderson, R. J., Losses in aluminium and aluminium-alloy melting, U. S. Bureau of Mines Reports of Investigations Serial No. 2239, April, 1921.
34. Anderson, R. J., and Capps, J. H., Gases in aluminium furnaces and their analysis, *Chem. and Met. Eng.*, vol. 24, 1921, pp. 1019–1021.
35. Anderson, R. J., and Capps, J. H., Constitution of gas atmospheres in aluminium-alloy melting furnaces, *Chem. and Met. Eng.*, vol. 25, 1921, pp. 54–60.
36. Anderson, R. J., Casting losses in aluminium-foundry practice, Trans. Amer. Foundrymen's Assoc., vol. 29, 1921, pp. 459–489.
37. Hartman, M. L., Carborundum refractories in non-ferrous metallurgy, Trans. Amer. Foundrymen's Assoc., vol. 29, 1921, pp. 524–530.
38. Anderson, R. J., Iron-pot melting practice for aluminium alloys, *The Metal Ind.*, vol. 19, 1921, pp. 189–190; 246–247; 285–287; 360–362; and *idem*, vol. 20, 1922, pp. 60–61; 309–311.
39. Anderson, R. J., Aluminium melting practice, *The Foundry*, vol. 50, 1922, pp. 737–741; 792–797; 823–826; 866–870; 919–924.
40. Anderson, R. J., Inclusions in aluminium-alloy castings, U. S. Bureau of Mines Tech. Paper 290, June, 1922.
41. Rosenhain, W., and Grogan, J. D., The effects of over-heating and repeated melting on aluminium, *Jour. Inst. of Metals*, vol. 28, 1922, pp. 197–207.
42. Czochralski, J., Die Löslichkeit von Gasen in Aluminium, *Zeit. fur Metallkunde*, vol. 14, 1922, pp. 277–285.
43. Gillett, H. W., and Mack, E. L., Electric brass furnace practice, U. S. Bureau of Mines Bull. 202, July, 1922, 334 pp.
44. Anderson, R. J., Preparation of light aluminium-copper casting alloys, U. S. Bureau of Mines Tech. Paper 287, October, 1922, 44 pp.
45. Lange, J. A., Melting aluminium for rolling into sheet, Trans. Amer. Foundrymen's Assoc., vol. 30, 1923, pp. 551–559.

46. Anderson, R. J., Aluminium and aluminium-alloy melting furnaces, Trans. Amer. Foundrymen's Assoc., vol. 30, 1923, pp. 562–604.
47. Anderson, R. J., Metallurgical requirements of refractories for use in the aluminium industry, Jour. Amer. Ceramic Soc., vol. 6, 1923, pp. 1090–1093.
48. Rosenhain, W., and Archbutt, S. L., The use of fluxes in the melting of aluminium and its alloys, The Metal Ind. (London), vol. 24, 1924, pp. 419–421; abst. of paper before The Faraday Society.

CHAPTER X.

PRODUCTION OF SECONDARY ALUMINIUM AND ALUMINIUM ALLOYS.

SECONDARY aluminium is produced by re-melting aluminium scraps and is used largely for re-melting in foundry practice for making alloys, although some secondary metal is employed for the production of aluminium rolling ingots. Secondary aluminium alloys are made by re-melting aluminium and aluminium-alloy scraps, and such alloys are used largely for re-melting in foundry practice. The production of secondary aluminium and aluminium alloys has grown to be a large branch of the aluminium industry. In the re-melting of aluminium-bearing scraps no refining can be done as in the case of copper, and secondary aluminum and aluminium alloys are normally less pure than the corresponding primary materials. The usual impurities found in secondary aluminium include copper, iron, silicon, manganese, and zinc; while secondary aluminium-alloy pig, say No. 12 alloy, may contain iron, silicon, manganese, magnesium, tin, and zinc, in addition to copper. Practice employed for the recovery of aluminium-bearing scraps is very variable; many types of furnaces are employed; and different fluxes are used. There is really no standardized practice, largely because of the variety of materials handled, and further, because so many small firms are engaged in the business.

In the United States, the smelting of secondary aluminium and aluminium alloys is carried out chiefly by companies engaged in the smelting and refining of non-ferrous metals and alloys in general, although there are a number of firms that specialize solely on aluminium and aluminium alloys. The literature relating to the smelting of scrap aluminium and aluminium alloys is scant, although many rules have been given in the trade journals for methods of melting and recovery. The principal investigation of the subject is that by Gillett and James,[18] reported in Bulletin 108 of the U. S. Bureau of Mines.

A few figures may be given at the outset to indicate the magnitude of the secondary aluminium industry in the United States. In the 10-year period 1913–1922, inclusive, the total amount of the secondary aluminium and aluminium alloys recovered in this country was 255,014,000 lb., valued at $91,295,320. Of this amount, there were recovered 116,312,000 lb. of aluminium as such, and 138,702,000 lb. of aluminium alloys. In 1913, according to estimates by the U. S. Geological Survey, 2,198 tons or 4,396,000 lb. of secondary aluminium, and 2,456 tons or 4,912,000 lb. of secondary aluminium alloys were recovered, mostly from clippings and borings. The entire recovered metal had a value of $2,199,480. In 1918, according to estimates by the U. S. Geological Survey,[20] the recovery of secondary aluminium was 6,050 tons or 12,100,000 lb., and of secondary aluminium alloys was 9,000 tons or 18,000,000 lb. The total recovery of secondary aluminium, as such, or in the form of alloys, amounted to 15,050 tons, or 30,100,000 lb., valued at $10,113,600. The total recovery in 1923 was the largest in the history of the industry, this amounting to 42,600,000 lb., valued at $10,824,600.

The bulk of the recovery of secondary metal and alloys is made from fabricating scrap, borings from machining, and automotive junk, but a fair amount is obtained from running drosses. The amount of available borings may be considered. Of a total production of 100,000,000 lb. of aluminium-alloy castings in 1923, about 92 per cent was for the automotive industry where the parts are subjected to considerable machining. The percentage by weight machined off from a rough aluminium-alloy casting, according to data obtained by Gillett and James [18] is about 15 per cent, on the average, and Table 80 gives some figures for the amount of material machined off different kinds of automotive castings. Since crankcases and transmission housings make up the greater part of the weight of aluminium-alloy castings in a motor car, the figures obtained as to these will have more effect on the total than those given by various small castings. According to the makers of a motor car in the $2,0co class, in which 47 aluminium-alloy castings are used, the total weight of the rough castings for each car is 166.19 lb., and the loss per car in borings owing to machining is 25.15 lb., or 15.1 per cent. On the basis that 15 per cent of the metal is machined off, and that 100,000,000 lb. of castings are produced, the annual production of borings is at the rate of 15,000,000 lb. This refers to 1923. If 20 to 30 per cent of the metal of the borings is lost on melting, then 3,000,000 to 4,500,000 lb. of aluminium alloys are lost (1923 basis). With the alloys selling at about 22 cents per lb., the total monetary loss was $660,000 to $990,000. No account is taken here of the losses arising in running down foundry wastes, but it may be estimated that the total monetary loss in smelting borings and foundry wastes is at the rate of $750,000 per annum.

TABLE 80.—*Percentage of metal machined off in finishing a variety of aluminium-alloy automotive castings.*[a]

Kind of casting.	Weight of rough casting, lb.	Weight of finished casting, lb.	Metal machined off, per cent.
Gear and transmission case	28.50	24.50	14
Small double-flanged exhaust elbows	0.42	0.35	15
Intake pipe for 6-cylinder motor	5.00	4.80	4
Gear case for small motor	8.90	7.00	21
Gear case for very small motor	1.58	1.50	5
Crankcase for 8-cylinder motor	94.00	80.00	14.5

[a] According to Gillett and James.

Secondary metal, selling at a lower price than primary aluminium pig, displaces an equivalent amount of new metal, and it accordingly must be considered by primary producers. Of rather greater importance to the consumer of aluminium manufactures, especially in the United States, where the production of primary aluminium is in the hands of one company, is the fact that the use of increasing amounts of secondary metal helps to lower the price.

In this book it is the object of the author to afford an insight into the general problems involved in secondary smelting, and detailed information as to certain aspects of the subject may be found in papers cited in the appended bibliography.

Production of Secondary Aluminium

CHARACTER AND QUALITY OF ALUMINIFEROUS SCRAPS.

There are a great variety of high aluminium-bearing scraps available for secondary recovery, and these may consist of aluminium scraps or aluminium-alloy scraps; it is desirable at the outset to differentiate between the two. Aluminium scraps may consist of aluminium drosses, skimmings, furnace splashings and drippings, as well as various kinds of scrap pieces resulting from manufacturing operations on substantially pure aluminium. Aluminium-alloy scraps may consist of aluminium-alloy drosses, skimmings, furnace splashings and drippings, various kinds of scrap pieces resulting from manufacturing operations on aluminium alloys, as well as foundry scrap, including defective castings, gates, risers, and runners. The various forms of aluminium scrap and aluminium-alloy scrap are discussed briefly below.

Aluminium Dross and Skimmings.—Aluminium dross may consist of aluminium oxide, Al_2O_3, plus metallic aluminium, in varying proportions, mechanically entangled therewith. Aluminium dross is normally the scum or oxidation product removed from liquid charges of substantially pure aluminium by skimming. The percentage of contained metallics is a function of the flux used in melting and the care with which the skimming is done. Aluminium dross of varying grade arises also in smelting operations on aluminium scraps. If the dross arising from such smelting operations is skimmed into water, it may contain a small percentage of mechanically entangled aluminium in the form a minute shotted globules. If cooled in air, the dross will contain no metallic aluminium. Aluminium dross normally contains aluminium nitride, possibly some aluminium carbide, silica, and other impurities. The terms aluminium dross and aluminium skimmings are used interchangeably. The aluminium content of dross as free metal may vary from 5 to 60 per cent, or more.

Aluminium Furnace Scrap.—Aluminium furnace scrap includes splashings from tapping and pouring, and drippings and skimmings from crucibles in melting practice for the production of rolling ingots. Such scrap is returned normally to the furnace, and not much of it is available for recovery by secondary smelters.

Aluminium Rolling-Mill Scrap.—Aluminium rolling-mill scrap is made up of slab and sheet shearings; buckled slabs and sheets that have caught in the mills during rolling; defective sheets, such as those rejected because of occluded dross, stains, splinters, and for other reasons; sheet shearings and trimmings; and various other scraps that arise in the production of sheets, shapes, rods, wire, etc. Practically all rolling-mill scrap, as such, is returned to the furnace room and re-melted, forming part of the regular charges with primary metal. Little of such scrap is available to secondary smelters, although a considerable percentage of the scrap aluminium made in the United States comes in the manufacture of aluminium sheets,[23] and in other rolled products.

Aluminium Fabricating Scrap.—Turning now to the kinds of scrap aluminium arising in fabricating operations on aluminium sheet, and other semi-finished forms of aluminium, this scrap may be termed fabricating scrap. In manufacturing operations on aluminium cooking utensils, various kinds of scraps result, including shearings, clippings, and punchings from punch-press operations, chips, coarse and fine "hay" from beading and trimming operations, stampings, and the like. Much fine light scrap arises in the production of aluminium pressings, punchings, and stampings. Where an aluminium cooking-utensil plant is attached to a rolling mill, all fabricating scrap is returned to the melting room either as such, or else is baled and then returned for melting. In the majority of utensil plants and stamping factories in the United States, however, in which fabricating operations are conducted on aluminium sheet purchased from outside sources, all fabricating scrap is sold in the open market or on contract to a secondary smelter. Scrap aluminium wire and cable is fairly pure, since the purest metal obtainable is required for electrical purposes. Bare wire and cable is desirable scrap for re-melting, but insulated wire is difficult to handle. There is growing use for aluminium wire of small diameter which is electrolytically insulated with a film of aluminium oxide,[13] and scrapped wire of this kind is difficult to re-melt.

Aluminium-Alloy Dross and Skimmings.—Aluminium-alloy dross consists of aluminium oxide, Al_2O_3, plus other metallic oxides, such as zinc oxide and copper oxide, plus metallic aluminium alloy, in varying percentages, mechanically entangled therewith. Aluminium-alloy dross is the scum or oxidation product removed from liquid charges of aluminium alloys by skimming. The constitution of aluminium-alloy dross depends upon the chemical composition of the alloy from which it is skimmed, and the general remarks made above as to aluminium dross apply to aluminium-alloy dross. Usually, aluminium-alloy drosses are sold in the open metal market for whatever price can be obtained, or they may be disposed of to smelters on contract. As in the case of aluminium skimmings above, the terms aluminium-alloy dross and aluminium-alloy skimmings are employed interchangeably. The aluminium-alloy content of skimmings may vary from, say, 5 to 60 per cent, depending upon the care used in skimming the charges, but the average content of free metallics may be taken as about 40 per cent.

Aluminium-Alloy Furnace Scrap.—This scrap includes splashings from ladling and pouring at the furnaces, and drippings and skimmings from crucibles or pouring ladles, in the melting rooms of foundries. Some aluminium-alloy furnace scrap arises in the production of aluminium-alloy rolling ingots, but the amount is small. Aluminium-alloy furnace scrap, including floor sweepings, is usually returned to the furnace.

Aluminium-Alloy Foundry Scrap.—Aluminium-alloy foundry scrap consists of gates, runners, risers, and defective castings incurred in the production of aluminium-alloy sand castings. Such scrap is normally returned to the melting room, and but little of it finds its way to the hands of secondary smelters.

Aluminium-Alloy Machining Scrap.—Most of the aluminium-alloy scrap available for secondary smelting consists of chips, borings, turnings, and drillings arising in machining operations on aluminium-alloy castings. A small amount of chippings and grindings is produced in cleaning rough castings in the cleaning rooms of foundries; the chippings are often returned to the melting room, while the fine grindings are generally sold to smelters. Since the greater part of the aluminium-alloy sand castings made is used for parts of motor cars, borings from the machine-shop of a motor-car plant may be regarded as typical. In drilling, milling, turning, and other machine-tool operations on aluminium-alloy castings, it is general practice to employ a lubricant such as a light cutting oil (kerosene) or emulsions of oil in a soap solution. If the borings from machining operations are permitted to stand in the air when wetted with such emulsions, they become superficially oxidized and caked together. In any event, borings, chips, and drillings become covered with whatever lubricant is used, and when they are swept along the floor, dust and dirt stick to the lubricant and become mixed with the borings. If iron, steel, brass, or white metals are machined in the same room as the aluminium alloys, there is danger that chips from the former will become mixed with the aluminium-alloy borings. In fact, it is the exception to find aluminium-alloy borings, as received from machine-shops, not contaminated with chips of other metals and alloys. There are great differences in the quality of aluminium-alloy borings, and such differences affect the recovery on smelting. Clean, large borings can be re-melted with fairly small losses, but small, dirty and oil-soaked borings give low recoveries.

Heavy Aluminium-Alloy Scrap.—Considerable heavy aluminium-alloy scrap arises in the junking of old motor cars. During 1919, the secondary metal market in the United States was flooded with great quantities of so-called airplane scrap, consisting of scrapped crankcases, oil pans, pistons, manifolds, and other parts from aviation engines. Heavy scrap of this character makes ideal material for re-melting in the production of secondary aluminium alloys, and automotive scrap is purchased by founders for direct melting to castings. Hence, there is competition between the smelters and the founders for this class of material. Foundry consumers can afford to pay higher prices for heavy automotive scrap than can the secondary smelters, and the use of this scrap is increasing. If, in round numbers, 1,000,000 motor cars are scrapped annually, of which at least half contain some 60 lb. of aluminium and aluminium alloys, then the amount of scrap arising as automotive junk is of the order of 30,000,000 lb. Probably half of this finds its way to smelters and the rest goes to foundries.

Miscellaneous Scrap.—The amount of miscellaneous scrap, including old aluminium wire, aluminium foil, discarded utensils and vessels, collapsible tubes, and small pressed and stamped articles bulks moderately large. Some of this scrap is lost and never recovered, while the remainder usually finds its way along with other scraps through the hands of junk dealers to the secondary smelters.

CONSTITUTION AND EVALUATION OF ALUMINIFEROUS SCRAPS.

Both aluminium drosses and aluminium-alloy drosses may be very variable as to chemical composition, and particularly as to their content of metallics; while in the case of borings the metal content may be considerably less than 100 per cent owing to admixture of oil, dirt, and other wastes. The amount of metallics in drosses may be very variable, from say 5 to 60 per cent, depending upon the origin of the dross, the flux used in melting, and the care employed in skimming. Not many published analyses have appeared giving the chemical constitution of aluminium-bearing drosses—possibly because of the sampling difficulties involved—and most of those given are incomplete. A typical analysis of an aluminium-alloy dross (dried sample) slagged off from a melting operation on light alloy borings was as follows: silica, 14.23 per cent; copper oxide, 4.47; iron oxide, 13.43; aluminium oxide, 64.70; and zinc oxide, 3.04 per cent (total 99.87 per cent). The material was screened before analysis to remove metallics, and as received, the dross contained 6.82 per cent volatile matter. Several methods have been devised for the wet chemical analysis of drosses, of which those by Bezzenberger [21] and by Hiller [22] may be mentioned. While the methods which have been recommended are entirely satisfactory for the chemical analysis of the actual oxidation product resulting from melting aluminium and its light alloys, the drosses and skimmings of commerce are not at all susceptible to any known sampling methods that will yield representative material for chemical analysis. In the case of material that contains practically no entangled metallic aluminium or aluminium alloy, and especially light fluffy drosses, as well as a mixture of dross and skimmings after long weathering, representative samples can be readily secured; but where there is much metal present, the fire assay is the only reliable method of evaluation for free metal content. In the purchase of drosses by smelters, the materials are bought on the basis of a fire assay, i.e., a miniature smelting run on a representative sample of sufficient size. Several runs, in which grab samples from different parts of the lot are taken, may be required.

Sampling and Assay of Drosses.—The sampling and evaluation of drosses is of much importance in secondary practice.

In sampling carload lots, a suitable method is as follows: A handful or more of material is taken from each wheelbarrow load removed from the car, until about 200 lb. or more are gathered. If there are large pieces in the lot, not only is a handful taken from each load, but one or more of such large pieces are taken also from each load, depending upon the proportion of large pieces in the shipment. The large sample resulting from this procedure is gone over, and the large pieces are broken down with a sledge hammer until the largest individual piece is not greater than $\frac{1}{2}$ in. diameter. About 200 gms. of this broken material is selected and sent to the chemical laboratory for evaluation and determination of chemical composition. About 200 lb. of the remainder is run through a swing-hammer pulverizer (cf. Fig. 109), screened, and then run over a magnetic separator, and the material after this treatment is run down in a pit furnace in order to obtain a check against the laboratory evaluation report. The 200-gm. sample sent to the laboratory is ground down in a mortar with pestle until the aluminium oxide and dirt adhering to the metal is loose and the largest pieces are about $\frac{1}{4}$ to $\frac{1}{2}$ in. diameter. Then, 100 gms., or about half of this ground sample, is taken and screened through a 30- or 40-mesh screen. The fines are discarded, and the material on the screen is gone over with a magnet, and the free-contained iron is determined on the basis of 100 gms. The metal content of the sample may be determined by running down the material left on the screen in a small graphite-clay crucible (No. 2 size) in a miniature gas-fired stationary furnace, e.g., Buffalo Dental type. The crucible is charged about two-thirds full, and the mass is puddled as it heats until the temperature is a cherry red. During this puddling operation the mass decreases in volume, and the remainder of the sample is added. When the whole charge has been worked up by puddling and raised to about 700° C., about one-half teaspoonful of cryolite is added as a flux, and the mass is puddled continuously until a thermit-like reaction begins. The crucible is removed then from the furnace, and the dross on the surface of the metal is skimmed off. The metal is poured into an iron pig mold, care being taken that no small particles of metal are lost by occlusion in the dross remaining in the crucible. The crucible is scraped cleanly to remove all metal and dross. The dross is quenched in air on a plate, and screened through a 30-mesh screen. The material remaining on the screen is returned to the crucible and re-heated, yielding additional metal. The total weight of metal obtained gives the metal content of the skimmings sampled, and the chemical composition of the metal is determined by analysis of drillings from the pig secured. The determination of the metal content of skimmings by the above method, as carried out in the laboratory, should check with the recovery obtained on smelting in the plant. This method is employed for evaluating both aluminium drosses and aluminium-alloy drosses. Following is the result of a determination made on a shipment of aluminium-alloy dross:

Constituent	Per cent
Metallic content	56
Magnetic iron	Nil
Copper	12
Iron	1
Zinc	40
Aluminium	Remainder

This means that a sample of skimmings contained 56 per cent metal and no free iron. The analysis of the alloy was 12 per cent copper, 1 per cent iron, 40 per cent zinc, and remainder aluminium.

Sampling and Assay of Borings.—In a general way, aluminium-alloy borings are sampled and evaluated according to the above described method for drosses and skimmings. With carload lots, a grab handful is taken from each wheelbarrow load as the shipment is unloaded, and the resultant large sample is thoroughly mixed. A sample of about 200 lb. is taken. From this, by quartering and mixing, about 200 gms. is taken for the laboratory run, and the remainder is used for the smelting test. In the laboratory run, half of the sample, or 100 gms., in treated with a magnet, and the free iron thus determined. The remainder is run down in a small crucible, as described above for drosses, and the metal content determined. The oil, if any is present, and the volatile matter may be determined on the other half of the laboratory sample, so that the recovery on the basis of the metal content of the borings may be calculated. The actual recovery on smelting borings may be expected to be a little higher than the recovery given by the laboratory run. A typical laboratory report on a shipment of borings is as follows:

Constituent	Per cent
Oil	3
Dirt	6
Magnetic iron	3
Metallic content	88
Copper	8.4
Iron	0.9
Zinc	0.4
Aluminium	Remainder

Evaluation of Scrap.—The evaluation or estimation of the recoverable metal in a lot of aluminiferous scrap, is made on the basis of the metal content and chemical composition of the metal obtained from the material to be purchased. The following gives an outline of the method of estimating price quotations on skimmings. It may be assumed that the laboratory report was as follows:

Constituent	Per cent
Metallic content	54
Magnetic iron	Nil
Copper	14
Iron	0.8
Zinc	40

This means that a given sample of skimmings contained 54 per cent free recoverable metal and no magnetic iron. The composition of the alloy is 14 per cent copper, 0.8 per cent iron, 40 per cent zinc, and remainder aluminium (45.2 per cent by difference). For purposes of calculation it may be assumed that it is desired to work the skimmings in with other metal so as to make a secondary casting alloy, using some of the copper, iron, and zinc, and all of the aluminium. The extra amounts of the copper, iron, and zinc above the percentages required to make, say, a modified No. 12 alloy, are of value only as scrap metals of the several kinds. If the modified No. 12 alloy is to contain 7 per cent copper, 0.8 per cent iron, 2 per cent zinc, and remainder (90.2 per cent) aluminium, the following calculations will obtain:

100 lb. of skimmings contains 54 per cent or 54 lb. of metal; 45.2 per cent of this 54 lb. of metal is aluminium for the No. 12 alloy, or 24.4 lb.; then

$\dfrac{24.4}{90.2} = 27$ lb. of No. 12 alloy.

The No. 12 alloy produced will contain the following amounts of the different metals:

Production of Secondary Aluminium

```
Copper.................    0.07 × 27 = 1.9 lb.
Iron...................    0.008 × 27 = 0.2 lb.
Zinc..................    0.02 × 27 = 0.5 lb.
Aluminium.............              24.4 lb.
                                    ───────
    Total..............              27.0 lb.
```

Now, 54 lb. of metal obtained from 100 lb. of skimmings contains the following amounts of the several metals:

```
Aluminium.............    0.452 × 54 = 24.4 lb.
Copper.................   0.14  × 54 =  7.6 lb.
Iron...................   0.008 × 54 =  0.4 lb.
Zinc..................    0.40  × 54 = 21.6 lb.
                                       ───────
    Total..............              54.0 lb.
```

The distribution of the metals for evaluation on the basis of 100 lb. of skimmings is given in Table 81.

TABLE 81.—*Distribution of metals for evaluation on the basis of 100 lb. of skimmings.*

Metal.	Metal in the skimmings, lb.	Metal in the No. 12 alloy, lb.	Metal to be evaluated, lb.
No. 12 alloy............	....	....	27.0
Copper.................	7.6	1.9	5.7
Iron...................	0.4	0.2	0.2
Zinc...................	21.6	0.5	21.1
Aluminium..............	24.4	24.4	....
Totals...............	54.0	27.0	54.0

The value of metals in dross or skimmings is determined by considering No. 12 alloy at the sales price and copper, iron, and zinc at their cost prices as scrap. The value of the lot of skimmings, therefore, is as given in Table 82.

TABLE 82.—*Value of lot of skimmings.*

Metal.	Cost per lb.	Metal in the skimmings, lb.	Value of the metal per 100 lb. of skimmings.
No. 12 alloy............	$0.30	27.0	$8.100
Copper.................	0.18	5.7	1.027
Iron...................	....	0.2	....
Zinc...................	0.06	21.1	1.267
Totals...............		54.0	10.394

The cost of handling plus overhead plus profit on the type of skimmings above is taken as 12 cents per lb. of metal produced. Hence, in order to smelt 100 lb. of the material it would cost $0.12 \times 54 = \$6.48$. The price that can be quoted on this material, then, is the difference between the value on selling and cost of producing, or

$$\begin{array}{r}\$10.394\\6.48\\\hline 3.914\end{array} \text{ per 100 lb.,}$$

or

$$\frac{\$3.914}{100} = \$0.03914 \text{ per lb.}$$

Under this condition, a quotation of 4 cents per lb. would be made.

The general plan of evaluation just described is applied to the evaluation of any type of scrap that is to be used in making up secondary alloys, and additional detailed calculations need not be given here.

FLUXES USED IN SECONDARY SMELTING.

While an almost endless number of fluxes have been suggested for use in smelting secondary aluminium materials, only a few of the more important fluxes can be considered here. The characteristics of the different fluxes vary markedly, depending upon their chemical composition, but, broadly, they may be divided into three classes, viz., (1) those that are used principally as liquid covers, and which probably reduce oxidation losses on melting; (2) those that may actually dissolve aluminium oxide, Al_2O_3, and thus induce coalescence of fine particles; and (3) those that are volatile at normal smelting temperatures, and which are useful principally because of their mechanical effects. Under the first class there may be mentioned sodium chloride and potassium chloride; under the second class there are included the alkaline fluorides and certain alkaline double fluorides, as well as various complex mixtures of salts; while under the third class there may be mentioned zinc chloride and ammonium chloride. Some of the various fluxes that have been used or suggested will be taken up below, but it is desirable first to indicate the requirements of a suitable flux.

Requirements of a Flux.—The question of a suitable flux for smelting secondary aluminium materials would seem to be solved if one can be found possessing the following attributes: power for dissolving aluminium oxide; melting point, about 500 to 700° C.; non-deliquescent; and cheap. These items are not given in the order of their importance, but the first two and last are probably of equal rank—a flux failing in any one of these points would not be entirely satisfactory. A flux which satisfied these requirements, but which was deliquescent, might still be desirable. However, a non-deliquescent flux would be preferable. The power of a flux for reducing the surface tension of globular aluminium particles is important, but information in regard to this aspect of the smelting question is so meager that power of reducing surface tension need not be considered now as one of the essential requirements of a flux.

Numerous mixtures of alkali chlorides and alkali fluorides have a direct solvent action on aluminium oxide, Al_2O_3, and these fluxes will dissolve the film of aluminium oxide that covers the individual particles of metal. Good recoveries have been obtained in British practice by the use of so-called fluoride fluxes of low melting point, even in smelting aluminium powder, but such fluxes attack ordinary graphite-clay crucibles with avidity. In a reverberatory or open-flame barrel furnace, a suitable refractory lining might be used that would resist the attack of these fluxes. A typical mixture recommended for use as a flux may contain 60 parts potassium chloride, 20 parts lithium fluoride, 12 parts sodium chloride, and 8 parts potassium sulphate; a similar mixture is used, but cryolite is substituted for the potassium sulphate; and another flux calls for the use of fluorides of calcium, potassium, or boron in a mixture of chlorides of the alkali metals. These fluxes are covered by the Schoop welding-flux patents (cf. Chapter XVIII). A flux should be non-deliquescent, since the addition of a water-containing flux to a liquid heat is dangerous. Moreover, salts without water of crystallization, i.e., anhydrous salts, are preferable. The melting point should be in the vicinity of the melting point of aluminium or lower, preferably about 500 to 600° C., so that it will be thoroughly fluid at the smelting temperature. The salts used in a flux should be readily friable, and most of them are. The question of density of the flux is important. If the flux is to be used merely as a liquid cover, the density should be considerably less than 2.70, but when used for the purpose of dissolving aluminium oxide, the flux should be about the same density as the material smelted, so that it will not segregate from the mass on initial smelting, if the melting is to be conducted in two stages. If the smelting is to be conducted in one stage, the flux should be of lower density than the material melted, so that it will rise readily to the surface of the bath and not become admixed with the metal. In this case, it is necessary, of course, to insure thorough contact of the flux with the metal by some mechanical means such as puddling. With two-stage smelting, a flux having practically the same, or lower, density as the material treated might be used first. This would be mixed with, say, borings, and charged. A flux of the same density as the alloy will not rise to the top, but will remain *in situ* in the charge. This is desirable, since what is wanted is the dissolution of the film of aluminium oxide from each particle of metal. On smelting, the resultant slag might be made less dense by charging a salt of lower specific gravity, which would combine with the slag and lower its density sufficiently so that the slag would rise readily to the top.

List of Fluxes.—Gillett and James [18] have discussed the question of fluxes at length and have given a detailed list of fluxes for smelting. Among other materials, aluminium chloride has been recommended, as has ammonium chloride; both these salts are volatile at moderate temperatures. Many binary mixtures of salts have been suggested or used, and there are a number of possible mixtures of alkali and alkaline-earth chlorides and fluorides having relatively low melting points that might be used as liquid covers or fluxes in smelting aluminium-bearing materials. Among others, the following mixtures may be mentioned: 85 : 15 calcium chloride-potassium chloride; 40 : 60 calcium chloride-potassium chloride; 85 : 15 calcium chloride-calcium fluoride; 36 : 64 sodium fluoride-aluminium fluoride; and 60 : 40 potassium chloride-potassium fluoride. All of these mixtures have melting points below 700° C. These mixtures have not been employed to any extent because they are too expensive, but the 85 : 15 sodium chloride-calcium fluoride mixture is used considerably, and this flux merits especial attention since it is a fluoride flux, so-called, that is cheap. This mixture melts at 785° C. and is very fluid at 800° C. While fluxes of high calcium-chloride content are fairly cheap, they are very viscous at temperatures far above their melting points, and they are also hygroscopic. Hence, they are not so good as the 85 : 15 sodium chloride-calcium fluoride mixture. Borax has been suggested and actually used to a slight extent as a flux in secondary smelting, as has calcium chloride. A great many complex mixtures (containing three or more salts) have been suggested and used, and a few of these may be cited, viz., 10 parts sodium carbonate, 2 parts potassium carbonate, 2 parts cryolite, and 1 part borax, used in the proportion of 1.5 lb. of flux to 100 lb. of borings; 16 parts calcium fluoride, 28 parts cryolite, and 56 parts aluminium fluoride; and equal parts of lithium chloride, potassium chloride, and sodium fluoride. Cryolite is used at times as a flux in secondary work. Calcium fluoride (fluorspar) is a good solvent for aluminium oxide, as is cryolite, and the former has been employed either alone, or in admixture with other salts, e.g., sodium chloride. Calcium fluoride is a good solvent for silicates, and the dirt in borings consists largely of various kinds of silicates.

In addition to calcium fluoride, certain other metallic fluorides, notably alkali and alkaline-earth fluorides, are solvents for metallic oxides. Normally, such metallic fluorides are not used alone as fluxes, but rather in combination with other salts to form complex mixtures. Weber[*] has suggested the use of fluorides of copper, nickel, or zinc, mixed with alkali fluorides. Puschin and Baskow,[11] and others, have examined several of the binary fluoride systems, e.g., potassium fluoride-aluminium fluoride, and lithium fluoride-aluminium fluoride. As a rule, metallic fluorides, principally alkali and alkaline-earth fluorides, are used in fluxes as constituents in binary or complex salt mixtures. Sodium chloride has been used for fluxing in smelting secondary aluminium alloys, either alone or mixed with fluorspar or other salts. Many mixtures used for welding fluxes (cf. Chapter XVIII) are metallurgically suitable for secondary smelting, since they dissolve aluminium oxide, but their cost makes their use prohibitive for the recovery of aluminium-bear-

---
[*] German Pat. No. 242,347, Dec. 30, 1910.

ing scraps. Modern welding fluxes are based on the experiments of Schoop, who demonstrated that aluminium oxide is soluble in certain alkali fluorides and mixtures with chlorides. Zinc chloride is often used as a flux in the smelting of borings, drosses, and skimmings in secondary practice, and it is also employed quite widely in ordinary foundry melting practice. While the value of zinc chloride as a flux in melting aluminium and its light alloys is debatable, there is good reason to believe that it aids in the separation of the dross from the metal. The action of zinc chloride is evidently largely mechanical, and the separation of dross from the metal bath is probably owing principally to the stirring action and upward currents caused in both by the volatilization of the salt. Considerable information on the melting points and character of fluxes are given in the report by Gillett and James,[18] and in papers by Plata,[1] Arndt and Loewenstein,[4] Żemczużny and Rambach,[6] Fedotieff and Iljinski,[10] Lorenz, Jabs, and Eitel,[12] and others. The discussion of the requirements of the aluminium reduction-cell bath in Chapter III may also be referred to in this connection.

COALESCENCE OF LIQUID ALUMINIUM GLOBULES.

The oxidation of aluminium, and its control, is one of the most important problems in secondary smelting practice, and the general aspects of oxidation have been discussed in Chapter IV. The affinity of aluminium for oxygen is exceedingly great, and this accounts for the low recoveries in running down fine scrap aluminium. In studying the problem of smelting fine aluminium-bearing scraps, two properties of aluminium must be considered, viz., (1) the ease of oxidation of the metal; and (2) the difficulty with which small globules of liquid aluminium coalesce, owing to the coating of aluminium oxide (or a mixture of oxide and nitride, or possibly carbide; or with fine dirt) or to surface tension, or both. Gillett and James [18] point out that if a mass of coated globules be pictured, such a mass would reveal a honeycomb structure, wherein the drops of liquid metal would be the honey and the film of aluminium oxide and dirt would be the comb. Expressed in another way, there is an emulsion of a solid (the coating film) and a liquid (the metal). On smelting fine scraps, if this emulsion is not broken up entirely, and the separate particles of metal freed from the enclosing coating, then the whole mass of aluminium oxide and dirt, together with the enclosed metal, will be removed on skimming. When air comes into free contact with such a hot porous mass, as it will when dross is skimmed, the minute liquid globules entangled in the mass oxidize rapidly. The heat evolved is so intense that a layer of such dross a few inches thick, placed on an iron plate $\frac{1}{4}$ in. thick, will burn a hole rapidly through the plate.

Of course, fairly large borings, free from dirt, form fairly large globules on melting, and such globules can break through the enclosing film of aluminium oxide by their own weight. Hence, they will coalesce fairly readily and melt down without much loss. In the case of fine borings, e.g., those that will pass a 20-mesh screen and that may have a thickness of only 0.005 in. or less, the situation is different. These small chips and borings form globules of almost microscopic size, and their weight is insufficient to rupture the enclosing film of aluminium oxide, much less a film of heavy dirt. Skinner and Chubb [13] have pointed out that the oxide film obtained in the electrolytic insulation of aluminium wire, which is from 0.0001 to 0.0004 in. thick, will stand up without rupture, in small coils, even when the coil is carrying so much current that the wire itself within this extremely thin shell is liquid. The fact that liquid aluminium particles fail to coalesce when in contact is analogous to the behavior of other metals, e.g., mercury, tin, and zinc, and this has been discussed detail by Gillett and James.[18]

METHODS SUGGESTED FOR THE RECOVERY OF SCRAPS.

Various methods for the recovery of aluminium-bearing scraps are suggested in the periodical and patent literature, but it is not possible to review all these methods in detail in the present book. Many references are given in the trade journals, and Gillett and James [18] have reviewed the literature in detail. The subject may be pursued at length by consulting the references appended to this chapter. In general, the various methods suggested fall in the following classes:

1. The borings or other scraps are to be fed back into regular aluminium electrolytic cells, used for the reduction of alumina. It is claimed that this method upsets the normal working of the bath, and it is likely that this device is not entirely practicable.

2. A liquid flux cover is to be used for covering the scrap during melting. The flux should melt at or below the melting point of aluminium, and it is employed for the purpose of protecting the metal from contact with air and hot combustion products in ordinary melting furnaces.

3. An actual flux that will dissolve aluminium oxide and dirt (silicates), or not, is suggested for use.

4. Borings and other scraps are to be charged into a liquid heel, thus excluding air from the fine metal.

5. A vacuum or retorting-type furnace is to be used, thus excluding air during melting. The stored heat in the walls of a previously heated furnace is suggested also, the borings being charged into the heated furnace and the furnace then closed.

6. Borings, or other light scraps, are to be charged to a pit, iron-pot, or other furnace and puddled with a rod so as to promote coalescence of the particles; the temperature of heating is kept fairly low, and a flux may be used or not during the puddling.

7. Borings or other light scraps are to be briquetted for the purpose of effecting a bond between the separate particles, thus cutting down the surface exposed, and making coalescence take place readily. Briquetting reduces the amount of air held on the surface of the borings, and permits easy submersion of the borings under the surface of a liquid heel.

8. It is suggested also that borings be subjected to pressure during melting so as to break down the film of aluminium oxide and thus promote coalescence. Or, a mass of borings may be subjected both to slight pressure and to constant stirring, as in the puddling method.

9. Cleaning the surfaces of borings by the employment of chemical reagents prior to melting is suggested.

10. The use of volatile fluxes, so-called, for promoting coalescence is another method. Volatile salts such as ammonium chloride, aluminium chloride, and zinc chloride which vaporize at 350° C., 180° C., and 730° C., respectively, are to be stirred into a mass of melted scrap consisting of globules of metal mixed with aluminium oxide and dirt (silicates). The gas evolved raises the particles, and has the result of mechanical stirring. Thus, some of the small liquid globules come into actual metallic contact with each other, and coalescence follows. It is also possible that nas-

cent chlorine or hydrochloric acid formed by the dissociation of the above chlorides may act upon the aluminium-oxide film, forming aluminium chloride, and cleaning the surfaces of the globules.

PREPARATION OF ALUMINIFEROUS SCRAPS FOR SMELTING.

Scrap castings constitute a rather large source of secondary aluminium alloys for smelting, but the disposal of this material in practice is very variable. A lot of scrap castings consists usually of a heterogeneous mixture of aluminium alloys of various compositions, such as junk dealers accumulate from the demolishment of old motor cars and from other sources. Some of this material is sold directly to founders while part of it goes to secondary smelters. On the whole, founders will do well to avoid the direct use of miscellaneous scrap, but heavy automotive scrap of known composition is quite suitable for general castings. With miscellaneous scrap, unless a foundry is equipped to make large heats of 1,000 lb. or more, which can be pigged, analyzed, and made up into definite alloys, it is inadvisable for a founder to look to this source for casting material. Both borings and drosses may be, and are, smelted as such, i.e., in the condition as received, but more often some preliminary preparatory treatment is given to such scraps. Both clean borings and other aluminium-alloy scraps can be re-melted and the metal recovered with comparatively small oxidation losses, but if small amounts of dirt and other foreign matter of a similar character are present, the difficulty of causing the separate particles to coalesce is greatly increased. In the case of drosses and skimmings, the greater the amount of aluminium oxide present the greater the loss on smelting, particularly if the temperatures are permitted to become high. Although considerable experimental attention has been given to the treatment of borings and drosses prior to melting, with the object in view of increasing the recovery of metal in commercial practice, it is customary to use only the more simple forms of treatment. Thus, drosses are normally crushed and screened, while borings are simply run over a magnetic separator before smelting.

Preparation of Drosses.—The smelting of drosses and skimmings is much more difficult and complicated than the mere re-melting of scrap castings, spatters, drippings, and the like. In practice, a lot of skimmings as received may consist of material varying from large spatters of metal to crumbly material consisting mainly of aluminium oxide 200-mesh in size and under. In order to run drosses and skimmings successfully it is necessary to make preliminary treatment of the material so as to remove as much of the non-metallics as is feasible. Owing

FIG. 109.—*Sectional view of swing-hammer pulverizer (The Jeffrey Manufacturing Co.).*

to the physical characteristics of drosses it is necessary to crush the materials by some form of machine that will jar loose the oxide. Crushing by rolling or pressing forces seems to knead the non-metallic part into the metallic part instead of freeing it. Jaw and gyratory crushers and rolls are totally useless for crushing drosses and skimmings. A swing-hammer pulverizer is the most suitable type of crusher for the preliminary crushing treatment of dross and skimmings. Fig. 109 shows a sectional view of a swing-hammer pulverizer. It is important that no tramp iron or large pieces of metallic aluminium be charged to the pulverizer, because such material may not only clog the bars but also cause breakage of the hammers and other parts; drosses and skimmings should be run over a picking belt and grizzly for the purpose of removing these materials.

After crushing, the dross should be screened, and the most satisfactory type of apparatus for this work is some type of a

Fig. 116.—*Hum-mer inclined screen; electro-magnetic vibration (The W. S. Tyler Co.)*.

mechanical or electrical vibrating screen. Aluminium oxide is hygroscopic, and this property of the material causes clogging of most other types of screens, e.g., revolving trommels, with resultant decrease in the efficiency of the separation. The openings in the screen may vary from 8 to 65 mesh, but this depends upon so many factors that it must be varied according to the material screened and the results required. The governing factor, of course, is the metallic content that is desired in the fines, and this depends upon the economic recovery of metal on smelting. Certain drosses can be extracted as to metal content very completely, while in the case of others, it is economical to leave a certain percentage of free metal in the fines. At times there is a market for fines of fairly high free metal content, and it is profitable, therefore, to sell the fines as such rather than to try to remove the ultimate percentage of metal possible. Fig. 110 is a photograph showing an electrically-vibrated screen in use at several plants for screening drosses and borings. With regard to the removal of mechanically admixed iron from drosses and skimmings, this is best done by running the oversize from the screen over a magnetic separator. Any roller magnet

FIG. 111.—*Magnetic separator for removal of free iron from crushed dross (Dings Magnetic Separator Co.).*

will accomplish this successfully. Fig. 111 is a photograph showing a magnetic separator suitable for the treatment of drosses after crushing and screening. After the preliminary treatment described in the foregoing, the material is furnaced according to various methods employed by different smelters. When skimmings are handled in the dry state, they may be stored in bins after preliminary treatment until required for smelting. Skimmings are concentrated by wet tumbling in some plants. When handled in this way they must be smelted promptly, because, if left wet in the air they will oxidize and heat.

Preparation of Borings.—Aluminium-alloy borings are normally given some preliminary treatment before smelting, and in order to remove admixed iron and steel chips, the borings are generally run over a magnetic separator. For borings, the magnetic separator should be of the disc type, as shown in Fig. 112, so that the borings will be worked over thoroughly. Borings are normally very oily, and a roll separator will not work efficiently on this class of material since too much of the aluminium-alloy chips are carried away with the iron. The effect of entrained dirt upon the recovery in smelting borings has been mentioned. When the borings are oily the dirt is held firmly,

FIG. 112.—*Dings magnetic separator (Dings Magnetic Separator Co.).*

and the expedient of screening is quite useless. Non-oily borings may be screened profitably for the purpose of removing dirt, cement, pieces of brick, and related foreign matter that may be present. Gillett and James [18] and others have examined experimentally the effect of sizing borings upon the recovery. It appears to be practical to size borings, since the larger borings can be handled more readily so as to obtain a high recovery than can a mixture of fine and coarse borings. Of course, some disposal would have to be made of the fines. Some tests have been made also on the effect of drying borings on the resultant recovery. Thus, the borings collected at a motor-car plant, where a water-soluble cutting compound is used in machining, are dried in a centrifugal dryer, running at the rate of 600 r.p.m. The dryer takes a charge of about 175 lb., and the liquid content is reduced to 3 to 5 per cent in 3 or 4 mins. of centrifuging. No water is added for washing out the cutting compound. After the borings are centrifuged they are stored in sheds until a sufficient amount has been accumulated, say a carload, and they are then sent to the smelter. The average recovery on the dried borings is reported to be 85 to 86 per cent, whereas on similar wet borings the recovery is only 50 to 60 per cent.

As mentioned previously, clean borings may be re-melted with fairly low oxidation losses, and small amounts of contained dirt in fine borings retard the coalescence of the particles. Gillett and James [18] have examined carefully the effect of washing and cleaning dirty and oily borings. Washing oily borings with hot water is useless, and practically no oil is removed by this treatment. Gasoline is a suitable reagent for washing oily borings, in that it removes the oil to a large extent when applied to fine borings, but the cost is doubtless excessive. This is true also of carbon tetrachloride. Sodium hydroxide solution (0.5 per cent) has been employed experimentally, and this appears to be worth while, although it has not been put into practice.

Baling Aluminium Scrap.—It is an economical procedure and now standard practice in some plants to bale all light aluminium scrap prior to re-melting. In general, baling machines are installed in aluminium-rolling mills for the purpose of handling light scraps, and more especially in utensil plants where much baling scrap originates. Baling is done by hydraulic baling machines similar in design and operation to those used for handling tin-plate and sheet-steel scrap in steel practice. The loose scrap is charged into a mechanical baler that forms a compact bale of convenient size for handling. Fig. 113 shows a view of an hydraulic baler (Logemann hydraulic scrap-metal bundling press). Baling furnishes a metallurgical product from light scrap that is easily handled and melted, and baled scrap invariably yields much lower melting loss than loose scrap. Baling is recommended for handling light scraps such as accumulate from fabricating operations on aluminium sheet.

Briquetting of Borings.—Since fine and light aluminium and aluminium-alloy scraps yield markedly higher oxidation loss on melting than do similar heavy scraps, it has been suggested, and tried experimentally, that these light scraps, especially borings, should be briquetted prior to melting. While briquetting, as applied to aluminium-bearing wastes, has not been employed so far in the United States, it is possible that it would be economical under some conditions, especially when there is a

FIG. 113.—*Logemann hydraulic baler for handling aluminium scrap (Logemann Brothers Co.).*

great spread in the market price between light scraps and heavy scraps. Borings, chips, and drillings can not be baled, and if they are to be gathered up into convenient form for handling, and at the same time into such a form that the oxidation losses on melting will be low, briquetting immediately suggests itself. Briquets of aluminium-alloy borings, if formed under very heavy pressure, are nearly as dense as pig metal. Sperry[9] has recommended this method of handling borings, and he states that, " in the treatment of aluminium chips, this process is particularly important as this metal, more than any other commercial one, is difficult to treat in such a form. When briquetted, the melting becomes a simple operation, and the resulting metal would be worth using. Metal now made from aluminium chips is of the poorest quality." Hirsh [14] has found that in melting loose borings a charge was run down in 50 mins. in a crucible with a loss of 13.8 per cent, while in melting briquetted borings the loss was only 8.1 per cent, and the time of melting 30 mins. The

FIG. 114.—*Ronay briquetting press (Stillman).*

briquetting of aluminium-alloy boring has been studied experimentally by Stillman [19] and others. In the production of aluminium and aluminium-alloy briquets for melting, as well as briquets of other metals and alloys, it is generally advisable that no binder be used, unless a material can be employed which will serve also as a flux on melting. In metal briquetting in general it is best to make the briquet by the medium of pressure alone. Fig. 114 is a photograph of the Ronay press made by the General Briquetting Co., New York, N. Y. This press is satisfactory for aluminium-alloy borings. Detailed description of the operation of this press has been given by Stillman.[19] Briefly, the briquets are made by filling a mold with the loose borings and then compressing the material under hydraulic pressure by means of a direct-acting plunger.

Furnaces in Secondary Smelting.—Furnaces for secondary aluminium work have not been standardized, as pointed out in Chapter IX, and many different types are employed. In smelting aluminiferous scraps the following types of furnaces are employed, viz: stationary and tilting iron-pot furnaces, reverberatory (open-hearth) furnaces, pit furnaces, and open-flame tilting and rotating furnaces. Strictly speaking, furnaces used for running down borings, drosses, and other high aluminium-bearing scraps, should not be called " refining " furnaces, since no actual refining of the material can be accomplished on melting. That is to say, the usual impurities, with the exception of zinc, can not be removed by a melting operation. Zinc may be volatilized from zinc-containing aluminium on melting at a sufficiently high temperature—above 930° C. (the boiling point of zinc)—but this normally gives rise to excessive dross losses from the aluminium. In practice, however, it is unusual to attempt to reduce the amount of impurities in a batch of material other than by adding substantially pure aluminium to the melting charge. The various types of furnaces employed for melting aluminium and aluminium alloys in general have been used in secondary practice. Some experiments have been made on the electric furnace smelting of scrap, but no application has yet been made.

Important work has been done recently on the development of furnaces especially for secondary aluminium smelting, and larger furnaces are coming into use. The reverberatory furnace is finding increased favor, and the application of mechanical puddling has been worked out at one plant. Mechanical puddling is of great advantage since it eliminates much expensive hand labor at the furnaces.

PRACTICE IN SMELTING ALUMINIFEROUS SCRAPS.

While the preliminary treatment of aluminiferous scraps, particularly drosses, is important and essential, the furnace operation is the most important factor in the recovery of metal on smelting. As has been indicated, the actual practice employed by producers is variable and many methods for recovery have been suggested. In a general way, practice calls for heating the various scraps in different types of furnaces and with different fluxes, and the heated mass is puddled or poked in order to promote coalescence of the metal particles. In American practice, fluxes of different kinds are invariably used in smelting drosses and borings, and effort is made to hold the smelting temperature as low as possible in order to avoid heavy oxidation losses. At the present time the two principal problems in secondary smelting are increasing the recovery, and finding better fluxes. Another important problem before secondary smelters is the production of a satisfactory " casting aluminium " from borings and drosses without the addition of any primary metal. This has not been accomplished, nor has the manufacture of such material so that foundry charges may be made up exclusively from it without the addition of any primary metal. It should be said that good grade casting alloys are produced by some secondary smelters, but the larger founders do not care to cast specification alloys from secondary metal without using some primary.

Practice in Smelting Drosses.—All-aluminium drosses may be run directly into secondary aluminium pig, and all-aluminium-alloy drosses of given composition may be run directly into secondary aluminium-alloy pig. When smelting miscellaneous lots of aluminium and aluminium-alloy drosses, i.e., drosses made up of skimmings from various sources, extraction and pigging of the metal is recommended. After this, definite alloys can be made up readily. The smelting operation on drosses can be carried out in any type of furnace that lends itself readily to hand puddling, i.e., stationary and tilting iron-pot furnaces, pit crucible furnaces, and reverberatory furnaces. From the point of view of labor and fuel costs, and recovery, the reverberatory-type furnace, if correctly designed, gives the best results. However, large production—10,000 lb. of metal or more per 24 hrs.— is necessary for the utilization of this type of furnace. Skillful operation is essential for smelting in any type of furnace, and inexperienced labor yields exceptionally poor results. Recovery in the smelting of drosses depends upon proper puddling at the " reacting temperature," so-called, and upon subsequent han-

Production of Secondary Aluminium

dling of the skimmed dross in the slag buggy. The reacting temperature may be defined as the temperature at which the thermit-like reaction begins, and this is about 850° C. In practice, the reacting temperature is considered to be reached when the heated mass of dross begins to glow in spots, and it is difficult to determine pyrometrically the precise temperature at which this begins. The rate of production of pig metal in smelting dross is governed quite largely by the type and capacity of the furnace used, and other things being equal, a rapid-heating furnace which is entirely under control of the operator is best adapted to rapid production. Proper puddling of the smelting mass in the furnace and of the dross in the slag buggy is developed only by experience.

Reverting to the question of the reacting temperature, shortly before this point is reached the flux should be added and stirred in. In large-scale practice, the dross skimmed from the furnace heat should be delivered to a slag buggy for puddling. While the dross is being worked up in the buggy, cold dross should be added to lower the temperature of the mass. The metal collects in the bottom of the buggy, and the overlayer of dross is removed, spread out in a thin layer to air-quench, and the metal is pigged. When cold, the dross is screened, and the coarse material that still contains metal is returned to the furnaces. Every precaution should be taken to avoid overheating while smelting drosses. In iron-pot smelting, good recoveries can be secured by proper puddling practice, air quenching of the dross, and screening of the dross taken off the furnace—the latter being returned and giving a further yield of metal. In iron-pot practice, it is impossible to reach the reacting temperature without the consequent dissolution of much iron from the pots and also without costly pot failures. Fig. 115 is a flow-sheet of the operations required in the preliminary treatment and smelting of aluminiferous drosses and skimmings. Cryolite or the 85 : 15 sodium chloride-calcium fluoride flux are used considerably for fluxing during puddling in smelting drosses, and the recovered liquid metal may be fluxed with zinc chloride before running into pigs.

Reverberatory, and other open-flame, furnaces are coming into use for smelting drosses, and this type of furnace is gaining ground largely because of the rapid production which it makes

466   *Metallurgy of Aluminium and Aluminium Alooys*

possible. In the case of reverberatory furnaces it is desirable to puddle the charge, and a furnace at one plant requires the labor of eight men for puddling. The material is puddled through four doors on either side, and the recovered metal is run off from a tap hole. It is of interest here to point out one feature in the construction of reverberatory furnaces for smelting

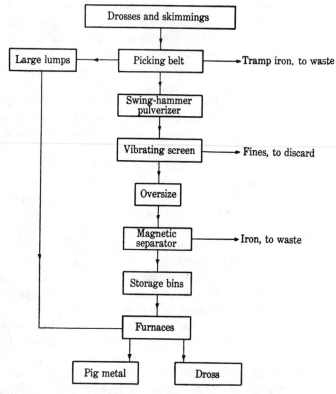

Fig. 115.—*Flow-sheet of operations in smelting drosses.*

aluminiferous scrap, viz., the preparation of a tight bottom. In the case of ordinary firebrick, or other refractory, bottoms, there is usually heavy leakage of the metal through the bottom. This occurs both in the ordinary melting of aluminium and its light alloys and in the smelting of scrap. The difficulty has been overcome at one plant by setting in a lining of wrought iron, in the form of an open box, and the refractory lining is laid inside the iron box in the usual manner.

Practice in Smelting Borings.—Practice in smelting aluminium-alloy borings is much simpler than that required in handling drosses, particularly as to preliminary treatment. Generally, borings are smelted as such, without preliminary treatment other than hand picking and magnetic separation, and the washing and sizing of borings has yet to be developed commercially. The method of puddling borings into a liquid heel is used commercially with fair results. On dirty, oily borings, with this method, recoveries as low as 75 per cent of the metal charged are common, and on clean borings recoveries as high as 96 per cent have been reported. In essence, the puddling method as carried out with a liquid heel for borings consists in the following, irrespective of the type of furnace used: A heel of liquid metal is heated to about 800° C., and a charge of borings is put in. The mass is puddled until a pasty state obtains, when the temperature is raised, more borings are charged, puddled, and so on until a charge is obtained that can no longer be worked effectively. At this stage, after again heating sufficiently, the flux is added, and the dross is skimmed, air-quenched, and screened. The oversize is returned to the furnace. Low yields in smelting borings are often due to improper fluxing and failure to recover the metal entrained in the dross on skimming. Fig. 116 shows a flow-sheet of the practice recommended for the preliminary treatment and the smelting of borings. In general, the safest way to handle borings, if received as mixed lots from various sources, is to run the borings into pigs and analyze the batch. Then, definite alloys can be made by subsequent re-melting. If borings are obtained in lots from one source, e.g., borings from No. 12-alloy automotive castings, they can be simply re-melted and pigged. When smelting either borings or drosses, the temperature should not be allowed to rise excessively, since oxidation losses will be greatly increased and poor recoveries secured. Moreover, if the temperature is allowed to rise very high, the recovered metal should not be poured at a high temperature, since the resultant pigs will have a poor surface appearance. In order to overcome this, the metal is cooled by adding some cold pig, and then run into molds. The fluxes used in smelting borings include cryolite, ammonium chloride, zinc chloride, sodium chloride, and the 85 : 15 sodium chloride-calcium fluoride mixture. In some plants, flux is added from time to time during the smelting in amount as deemed necessary, while in others

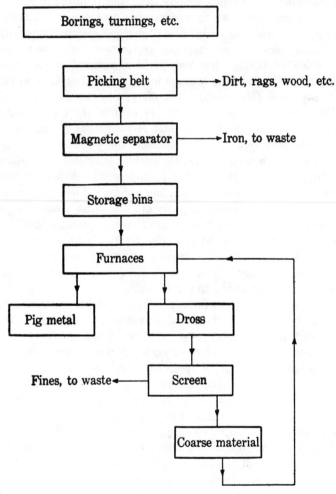

FIG. 116.—*Flow-sheet of operations in smelting borings.*

flux in definite amount from 10 to 30 per cent by weight of the borings is charged, depending upon the character of the borings. Borings are smelted in the United States in pit, iron-pot and reverberatory furnaces, and the use of the latter is gaining ground. Practice in smelting borings at several plants has been described in detail by Gillett and James.[18]

Blending of Scraps.—For economical and profitable handling of aluminiferous scraps in making secondary aluminium alloys, miscellaneous lots of scraps must be blended, and a sample blending chart is given in Table 83, which shows the amounts of the different materials used in a charge for the production of secondary No. 12-alloy pig at a smelting plant.* The specifications for this alloy at the plant in question are as follows: copper, 6.5 to 7.5 per cent; iron, less than 1 per cent, and zinc 1 to 2 per cent. With zinc permitted in the specifications, the plant can purchase all kinds of aluminiferous scrap and convert it into a salable product by proper blending of the scraps.

TABLE 83.—*Blending chart for charge in smelting for the production of secondary No. 12 alloy.*[a]

Material.	Weight, lb.	Weight, metal content, lb.
Skimmings	100	75
Borings	75	65
Re-melted alloy	39	39
50 : 50 copper-aluminium alloy	7.5	7.5
No. 1 aluminium clippings	15	15
Total	236.5	201.5

[a] Approximate composition of the resultant alloy: copper, 7 per cent; iron, less than 1 per cent; zinc, less than 1.5 per cent; and aluminium, remainder.

Unless a smelter runs borings on toll for a plant that produces a large quantity of chips, it is normally necessary to purchase quantities of different kinds of scraps, i.e., borings, drosses, scrap castings, etc., in the open metal market. The economical disposal of lots of miscellaneous scraps is a matter calling for the most skillful blending practice, if secondary pig of good casting qualities is to be produced. The problem confronting the average secondary smelter is the production of pig of suitable casting quality from a miscellaneous assortment of scrap, and much ingenuity is required in working in lots of scraps of widely variable chemical composition. Referring to Table 83, the re-melted metal is added after the skimmings and borings have

---
* Private communication, Sept. 30, 1921.

been run down. After this, the mass is heated but little, so that when it is ready to pour, the temperature is below 700° C. The intermediate alloy is added last. At the low temperature employed, the 50 : 50 copper-aluminium alloy must be used for introducing the copper. This agrees with the alloying practice employed in many foundries.

Recovery on Smelting.—The actual recovery of metal from aluminium-bearing scrap is very variable in practice, depending upon the kind and quality of the scrap. The recovery on aluminium drosses and aluminium-alloy drosses is always lower than on all-metallic scraps. Excluding heavy scrap, the recovery on aluminium-bearing scraps, such as borings or chips, small buttons from foundry-floor sweepings, grindings, and the like, may vary from 40 to 90 per cent of the metal charged. The actual recovery in the case of such material varies with the method of smelting, the furnace used, the flux employed, the size of the material smelted, and its cleanliness (freedom from oil, dirt, and other foreign matter) and other factors. Referring especially to ordinary dirty borings resulting from machining operations on automotive castings, made in 92 : 8 aluminium-copper alloy, it is doubtful whether the average recovery is more than 65 per cent in practice. One smelter claims * an average recovery of 88 per cent on such material, with recoveries running as high as 92 per cent at times. These figures seem rather out of line with general practice. The recovery obtainable on drosses containing 40 per cent metallics may be expected to be about 50 per cent on the average.

The actual recovery can determine whether a secondary smelting plant can operate or not. Gillett and James [18] have cited the experience of certain firms in this connection. One firm which was organized solely to run down borings into aluminium-alloy pig, shut down its plant after a few months, while another company ran its aluminium department only intermittently, finding that when aluminium was scarce it was advisable to stop re-melting borings because their price rose above that at which a profit could be made with the recovery secured. During such conditions in the metal market, the borings go to small refiners, or else to foundries in which borings are run down separately or added to the regular melting charges. In many

---
* Private communication, June 2, 1921.

instances, however, no cost records are kept by such plants, and the managers may be melting borings at considerable loss without being aware of the fact. According to one founder, with primary aluminium pig selling at 25 cents per lb. and copper at 15 cents, primary No. 12 alloy would sell at about $24\frac{1}{2}$ cents per lb., while No. 12-alloy borings would sell at $12\frac{1}{2}$ cents, and secondary pig made from such borings would sell at about 22 cents. If the cost of smelting, including the cost of fuel, labor, overhead, etc., is assumed to be 1 cent per lb. of secondary pig produced, it is evident that with a recovery of 50 per cent on the gross weight of borings, the cost of producing a pound of pig would be 25 cents for metal and 1 cent for smelting, or 26 cents. With secondary pig selling at 22 cents, this would mean a loss of 4 cents per lb. of pig. With a 70 per cent recovery the cost would be about 20 cents, or a net profit of 2 cents per lb. If the recovery were 60 per cent, there would be a slight loss, but with 80 per cent recovery the profit would be about 5 cents per lb. In these calculations it is assumed that the borings cost $12\frac{1}{2}$ cents. Gillett and James [18] have pointed out that if it be assumed that the average recovery from dirty borings and related scraps is 65 to 70 per cent on the basis of the metal content, and if 90 to 95 per cent recovery can be obtained on borings kept clean and properly melted, then 20 to 30 per cent of the metal machined off castings is unavoidably lost.

The cost of smelting aluminium-bearing wastes is very variable, from 1 to 10 cents per lb., depending upon the kind of material handled and the practice employed.

COMPOSITION AND QUALITY OF SECONDARY ALUMINIUM AND ALUMINIUM ALLOYS.

In general, the amounts of the usual impurities in both secondary aluminium and in secondary aluminium alloys are substantially greater than in the corresponding primary materials. This is to be expected, even in the case of secondary products obtained by a simple re-melting operation on clean, fairly heavy scraps, because impurities are introduced from the containing vessels used for melting. With regard to the behavior of secondary aluminium alloys, i.e., secondary No. 12 alloy, in foundry practice, it may be said, by and large, that this is a very moot question. Most foundrymen would hesitate to use all secondary No. 12 alloy made from borings in important castings, such as in automotive work, and the present requirements call for the addition of at last 30 per cent primary aluminium to secondary material in order to secure satisfactory castings. Some foundrymen feel that they must use at least 60 per cent of primary metal. On the other hand, some smelting men feel that 60 per cent is an unnecessarily large amount, and that the proportion might be 30 per cent primary metal and 70 per cent secondary metal. In the case of large castings of complicated design, such as crankcases for large motors, it is a question of considerable debate among founders as to whether the use of secondary pig metal in the charge increases the casting losses owing to cracking in the mold. It is true that in the case of some lots of secondary No. 12 alloy, the material has given low values in the hot-shortness test, which would indicate a tendency toward cracking in the mold. *Per contra*, certain foundrymen claim that a certain grade of secondary No. 12 alloy pig made from borings, used in amounts up to 30 per cent by weight of the charge, markedly decreases the loss from cracking, although other secondary pig of practically the same chemical composition increases the loss.

Such defects as cracks, draws, porosity, and brittleness in aluminium-alloy castings are often attributed to secondary materials. It is well known that large smelters with a desire to make the best possible product have such practice that the metal produced is of good quality and suitable for founding, but certain unscrupulous people have engaged in the secondary smelting business, with the result that there have been placed on the market certain brands of secondary products which can be used in foundry practice only with the greatest difficulty. Producers of scrap, i.e., machine-shops operating on aluminium-alloy castings, in general take little interest in keeping borings clean, i.e., free from iron, brass, bronze, white metals, etc., and better care in keeping foreign impurities out of borings would result in a better quality in the average secondary alloys. Aluminium and its light alloys have received much severe condemnation owing to the poor performance of inferior secondary material, and it is of the greatest importance that the secondary end of the business be placed on a strictly rational basis. Thus, certain castings may be made in an experimental way for some manufacturer, who may intend to substitute an aluminium alloy for brass, bronze, cast iron, or steel with the idea that the aluminium alloy will serve more advantageously than one of the other non-ferrous alloys or steel. If the castings are made of secondary materials, and if, owing to the poor quality of these materials they break down or do not serve the purpose which engineering data indicate they should, then it is concluded immediately that an aluminium alloy is unsuited for the purpose in question. Actually, the fault does not lie with the aluminium alloys, as such, but rather with inferior secondary products which have been manufactured by unscrupulous and ignorant makers. Secondary aluminium and aluminium alloys should not be condemned *per se*, since there are firms with established reputations that produce good casting alloys. So far as is known, however, secondary aluminuim and aluminium alloys are not suitable for die casting, but they are used successfully in sand and permanent-mold work.

The employment of secondary aluminium and aluminium alloys has increased rapidly in recent years, and any legitimate steps that can be taken to overcome trade objection in certain quarters to the use of good secondary metal will not only result in better business for the secondary smelters but will also lower the cost of aluminium manufactures to the ultimate consumer. There are many possible applications of aluminium that are now out of the question because of the high price of the metal. Of course, for some manufactures, e.g., aluminium wire in which special purity is required, it is quite necessary to employ only high-grade primary metal, since it is normally impossible for secondary aluminium to meet the specifications, but in alloy-castings production, as well as in sheet rolling, secondary metal can be used either entirely or in part. In fact, this use of secondary aluminium and aluminium alloys is rapidly increasing. A still greater amount of secondary aluminium could be used if the various scraps were handled more carefully and kept cleaner before reaching the smelter. Of course, as the quality of the products turned out by the smelter becomes better, trade objections will gradually disappear. In the production of aluminium-alloy sand castings, it is certainly poor business to use high-grade primary aluminium when a lower grade and less costly secondary metal will do just as well, and in many cases the primary metal is employed solely because precedent is followed and the founder thinks that he can not use secondary metal. In the foundry business, moreover, competition for orders is so keen that it is absolutely necessary to use secondary metal, in whole or in part, depending upon the job, and the simplest method that can be resorted to for reducing costs is to use scrap or secondary metal in greater amounts. It should, of course, be emphasized that the secondary or scrap metal to be used should be of suitable quality and the foundry practice should be of the best, since the indiscriminate employment of scrap and poor-grade secondary metal would be disastrous. It is possible to purchase secondary aluminium-alloy pig from reputable smelters that can be safely and fully substituted for primary aluminium in foundry practice, but in all fairness it should be said that it is also possible to purchase secondary metal from other firms that is entirely useless for casting.

There has been some secondary aluminium-alloy pig of exceedingly poor quality on the market, evidently made by running down a miscellaneous assortment of aluminium-alloy, babbitt and other white-metal, and brass and bronze borings. Such material resembles aluminium only in color, and it is useless for casting purposes. For example: a sample of so-called "casting aluminium" offered to a foundry by a smelter had the following composition: 17.4 per cent copper, 8.4 zinc, 1.8 iron, 2.4 silicon, 0.5 manganese, and 69.5 per cent aluminium (by difference). Table 84 gives some analyses of secondary aluminium and aluminium-alloy pig as marketed in the United States. Primary No. 12-alloy pig is made up so as to contain 7.0 to 8.5 per cent copper, no zinc, 0.5 to 0.7 iron, 0.25 to 0.40 per cent silicon, and remainder aluminium. When such pig is melted in iron-pot furnaces in foundry practice, at least 0.2 per cent iron is taken up from the pots, skimmers, and ladles, and about 0.05 per cent silicon from sand adhering to gates. Secondary No. 12-alloy pig and castings made from such pig are always higher in impurities than primary pig and castings made from such pig, respectively. One alloy made by a large secondary smelter contains about 8 per cent copper, 4 per cent zinc, and the remainder aluminium, plus the impurities iron and silicon. So-called "casting aluminium" is made generally by smelting an indiscriminate assortment of high aluminium scraps, and it may contain most of the usual non-ferrous metals in amounts varying from traces to several per cent.

TABLE 84.—*Composition of secondary aluminium and aluminium-alloy pig.*[a]

Material called.	Chemical composition, elements per cent.							
	Al [b]	Cu	Fe	Si	Zn	Mn	Sn	Pb
98–99 per cent aluminium pig....	97.72	0.42	0.82	0.30	0.74	....	....	....
Casting pig....................	90.40	7.23	1.08	0.21	0.99	....	....	....
Boring pig.....................	82.2	4.2	9.3	0.6	2.5	1.2	....	....
Boring pig.....................	87.3	7.7	2.8	1.9	0.3	....	....	....
No. 12-alloy pig................	90.7	7.8	0.7	0.4	0.4	....	....	....
Casting pig....................	84.4	6.8	0.7	0.2	6.0	....	0.6	1.3
No. 12-alloy pig................	89.9	7.5	1.3	0.3	1.0	....	....	....
No. 12-alloy pig................	88.6	8.8	1.4	0.4	0.8	....	....	....
No. 12-alloy pig................	90.4	6.6	1.3	0.2	1.5	....	....	....
Casting pig....................	64.1	6.4	1.0	0.3	25.0	....	1.3	1.9
Casting pig....................	80.35	3.9	3.0	0.5	12.1	0.15	....	....

[a] From various sources.
[b] Aluminium by difference.

Impurities in Secondary Pig.—The chemical composition of secondary aluminium and aluminium-alloy pig is very variable, depending upon the chemical composition and quality of the scrap materials smelted. Secondary aluminium made from clean sheet clippings is fairly pure, having aluminium content but slightly inferior to that of the clippings, if properly smelted. The chemical composition of secondary aluminium-alloy pig smelted from aluminium-alloy borings depends upon (1) the chemical composition of the castings from which the borings were machined; (2) the presence, in the aluminium-alloy borings, of chips or borings of other metals or alloys such as iron or steel, babbitt, white metals, brass, or bronze; (3) the presence of dirt, cement, small pieces of brick, or other hard foreign materials; (4) whether or not the borings have been washed or cleaned for removing dirt, or subjected to magnetic separation for removing iron; and (5) the kind of smelting process used, the furnace employed, and the temperature attained. White metal, babbitt, brass, or bronze borings, if present, can not be readily separated from aluminium-alloy borings, and thus give rise to impurities in the resultant pig. Copper, zinc, lead, tin, and antimony may come from these sources. The various impurities found usually in secondary aluminium and in secondary aluminium alloys are discussed briefly below.

Antimony may be found at times in lots of secondary aluminium alloys, being derived from babbits or other antimony containing chips in the borings. Antimony is an undesirable impurity. Copper is normally present in all secondary aluminium and aluminium alloys. Copper is present to the amount of 7 to 8.5 per cent in all No. 12-alloy borings, and it is present in borings obtained from machining most aluminium-alloy castings, including the zinc-containing alloys. From the foundry point of view it is desirable that the copper content of secondary No. 12 alloy should be about 7 to 8.5 per cent, although if it is too high or too low, aluminium or copper may be added to yield the correct composition in the foundry charges. However, in small foundries, to which no chemical laboratory is attached, difficulties may arise in casting practice because of too high or too low copper in the secondary pig. Both too high and too low copper content, say 15 per cent and 5 per cent copper, respectively, may give rise to cracks, draws, and other defects in the castings.

Iron is a normal impurity in all secondary aluminium and aluminium alloys. In the case of borings, iron or steel chips may be admixed in the machine-shop and the magnetic separation of iron from such contaminated borings may leave from 0.5 to 1 per cent of free iron in the borings, especially if they are very oily. Upon melting, the admixed iron or steel goes into solution in the aluminium, and varying amounts of iron are found in the resultant secondary alloys. In the iron-pot puddling process for the recovery of borings a certain amount of iron is dissolved from the iron pots. Of course, iron is a normal impurity in both primary aluminium and primary aluminium alloys, and upon melting the primary materials in iron pots in foundry practice, some iron is dissolved. Since the foundry scrap, including gates, risers, and defective castings, is returned to the melting room, the iron content of the castings becomes increasingly higher. Moreover, certain light aluminium-copper-iron alloys, notably 90.5 : 8 : 1.5 aluminium-copper-iron, are used for some castings. Aluminium-alloy die castings contain rather high percentages of iron, on the average, and often up to 3 per cent. Lead is found at times in secondary aluminium alloys, resulting from chips of white metals, leaded bronzes, and other lead-containing alloys in the borings. Lead is an undesirable constituent in aluminium alloys because it does not alloy with the aluminium but liquidates out on melting, and will be found mechanically enclosed in the form of small globules in the metal. If lead-containing secondary pig is melted and poured into castings the lead will tend to sink to the bottom of the mold, thus giving a weak and faulty casting. Lead should not be permitted even in small amounts in secondary aluminium alloys. Manganese is frequently found as an impurity in secondary aluminium and aluminium alloys, being derived largely from clippings of 98.5 : 1.5 aluminium-alloy sheet or from borings of aluminium-copper-manganese alloys. In amounts less than 0.5 per cent, manganese is apparently not harmful to certain aluminium-copper alloys and aluminium-copper-zinc alloys, but if present in too large percentages, it increases the shrinkage of the alloys, and thus gives rise to casting troubles in the foundry.

Silicon is a normal impurity in both primary and secondary aluminium and aluminium alloys, and on smelting aluminiferous wastes in graphite-clay crucibles or against refractory firebrick bottoms, silica is reduced to silicon and alloys with the aluminium. Silicon may also be reduced on re-melting gates to which molding sand adheres, or in melting borings that contain siliceous dirt. The reduction of silica by aluminium is not pronounced below 800° C., and even when melting in graphite-clay crucibles only a small amount of silicon is taken up, provided the temperature is kept low. With higher temperatures, however, the reduction of silica takes place more readily, so that at the temperatures attained, and necessary, in smelting borings, some increase in the silicon content is to be expected normally if the borings contain siliceous dirt. The use of a flux that will dissolve or soak up the dirt and consequently remove it from contact with the metal might be expected to decrease the amount of silicon reduced and taken up at any given temperature. So far as is known, silicon is not harmful to the mechanical or casting properties of aluminium and aluminium alloys, but rather improves them. The use of borings from aluminium-silicon and aluminium-copper-silicon alloys would, of course, introduce silicon into the resultant secondary pig. Tin is present in some secondary aluminium alloys, arising either from bronze chips, solder chips, babbitts, or other tin-containing pieces which may become admixed with the borings. It may also come from tin plate used for introducing iron into certain light aluminium-copper alloys. Some aluminium-copper-tin alloys are used for certain special castings, and when the borings from such castings are smelted together with other borings, tin will be found in the resultant alloy. In general, tin may be regarded as a harmless impurity in secondary No. 12 alloy, although it may often be accompanied by lead—both impurities resulting from admixed solder chips in the borings. The effect of tin in relatively large percentages upon No. 12 alloy is not appreciable; tin tends to soften the alloy and make it more ductile without markedly affecting the tensile strength. Further, as a matter of fact, ternary aluminium-copper-tin alloys appear to contain fewer blowholes on the average than aluminium-copper alloys, and they take a higher polish.

Zinc is often present in secondary so-called No. 12 alloy to the amount of 0.5 to 2 per cent, and it is invariably present in larger amounts, up to 8 per cent or more, in so-called "casting aluminium." Zinc in No. 12-alloy castings is frowned on by many purchasers, even if present to the extent of only traces to 0.5 per cent, in some cases because it is known that zinc can not be in No. 12 alloy made from primary metals, but more often because of a fancy that zinc-containing aluminium alloys are weak under shock stresses. As a matter of fact, the presence of small amounts of zinc in No. 12-alloy secondary pig is of little practical consequence in casting work. A mixture of aluminium-alloy borings, as bought by a smelter, will consist largely of No. 12-alloy borings, but such a mixture may contain also some borings from alloys containing 10 to 30 per cent zinc. Hence, the resultant secondary pig made from such a mixture may contain varying amounts of zinc. If a few tenths of 1 per cent of zinc are present in a casting it indicates that some scrap, or else secondary No. 12 alloy made in part from scrap, had been used. There is no definite evidence to show that a zinc content of up to 0.5 per cent has any detrimental effect upon the strength or endurance of No. 12-alloy castings.

The inter-related effects of the various impurities, e.g., zinc, tin, lead, iron, silicon, etc., in secondary No. 12 alloy upon the casting properties are in much doubt, and on this basis alone it is highly undesirable that so-called "casting aluminium" be used for important castings. Some fairly well-substantiated data are available as to the inter-related effects of copper, iron, and zinc, and these may be presented briefly here. Thus, 2 per cent of iron in No. 12 alloy gives rise to too brittle an alloy, but 0.75 to 1.25 per cent of iron does not appear to be detrimental. On the other hand, if the iron content rises much above 1 per cent, the copper content should be reduced correspondingly below 8 per cent in order to prevent brittleness, since iron acts much like copper in hardening and strengthening aluminium. While zinc forms solid solutions with aluminium up to about 18 per cent zinc, it appears that if zinc be added to No. 12 alloy, more of the $CuAl_2$-aluminium eutectic is thrown out than without zinc, so that in the useful ternary aluminium-copper-zinc alloys the copper is present in smaller amount as the zinc is present in larger amount. For example: the 91.5 : 7.5 : 1 aluminium-copper-zinc alloy has about the same strength when cast as the 92 : 8 aluminium-copper alloy. In an aluminium alloy containing 20 per cent of zinc, not more than 4 per cent of copper should be present, since otherwise the resultant alloys will be too brittle. By proper reduction in the copper content, the effect of high iron content and of a little zinc (introduced by the use of some secondary pig) can be neutralized largely, and the resultant alloy will still have closely the properties of No. 12 alloy as regards casting properties, shrinkage, tensile strength, and elongation.

SELECTED BIBLIOGRAPHY.

The literature dealing with secondary smelting of aluminium and its alloys is scant, but there are a number of published papers that give information bearing on the general subject. Some references on oxidation and fluxes are included in the selected list below. The references on the oxidation and nitridation of aluminium appended to Chapter IV may also be consulted.

1. Plata, W., Erstarrungserscheinungen an anorganischen Salzen und Salzgemischen, *Zeit. für phys. Chem.*, vol. 58, 1907, pp. 350–362.
2. Schoop, M. V., Autogenous welding of aluminium, *Electrochem. and Met. Ind.*, vol. 7, 1909, pp. 151–153.
3. Seligman, R., The welding of aluminium, *Jour. Inst. of Metals*, vol. 2, 1909, pp. 281–287.

4. Arndt, K., and Loewenstein, W., Über Lösungen von Kalk und Kieselsäure in geschmolzenem Chlorcalcium, *Zeit. für Elektrochem.*, vol. 15, 1909, pp. 784–790.
5. Sperry, E. S., Remelting of aluminium chips or borings, *Brass World*, vol. 6, 1910, p. 278.
6. Żemczużny, S., and Rambach, F., Schmelzen des alkali Chloride, *Zeit. für anorg. Chem.*, vol. 65, 1910, pp. 403–428.
7. Neumann, B., and Olsen, H., Production of aluminium as a laboratory experiment, *Met. and Chem. Eng.*, vol. 8, 1910, pp. 185–188.
8. Sperry, E. S., Briquetting metal chips, *Brass World*, vol. 7, 1911, p. 41.
9. Anon., Briquetting machine for metal borings, *Engineering*, vol. 94, 1912, pp. 737–739.
10. Fedotieff, P. P., and Iljinski, W., Beiträge zur Elektrometallurgie des Aluminiums, *Zeit. für anorg. Chem.*, vol. 80, 1913, pp. 113–154.
11. Puschin, N., and Baskow, A., Das Gleichgewicht in binären Systemen einiger Fluorbinden, *Zeit. für anorg. Chem.*, vol. 81, 1913, pp. 347–363.
12. Lorenz, R., Jabs, A., and Eitel, W., Beiträge zur Theorie der Aluminiumdarstellung, *Zeit. für anorg. Chem.*, vol. 83, 1913, pp. 39–50.
13. Skinner, C. E., and Chubb, L. W., The electrolytic insulation of aluminium wire, *Trans. Amer. Electrochem. Soc.*, vol. 26, 1914, pp. 137–147.
14. Hirsch, E. F., Metall-briketts, *Elektrochem. Zeit.*, vol. 35, 1914, pp. 1092–1094.
15. Pannell, E. V., Recent developments in aluminium; some notes on autogenous welding, *Trans. Amer. Inst. of Metals*, vol. 9, 1915, pp. 167–180.
16. Gillett, H. W., Melting aluminium chips, *Trans. Amer. Inst. of Metals*, vol. 9, 1915, pp. 205–210.
17. Coulson, J., Reclamation of magnalium from turnings, *Trans. Amer. Inst. of Metals*, vol. 9, 1915, pp. 336–342.
18. Gillett, H. W., and James, G. M., Melting aluminium chips, U. S. Bureau of Mines Bull. 108, August, 1916, 88 pp.
19. Stillman, A. L., Briquetting of non-ferrous light metal scrap, *The Metal Ind.*, vol. 15, 1917, pp. 526–529.
20. Dunlop, J. P., Secondary metals in 1918, Mineral Resources of the United States, 1918, part 1, p. 602; and see also preceding and later volumes.
21. Bezzenberger, F. K., The evaluation of aluminium dross, *Jour. Ind. and Eng. Chem.*, vol. 12, 1920, pp. 78–79.
22. Hiller, H., Über die Analyse von Aluminiumasche, *Zeit. für angew. Chem.*, vol. 33, 1920, pp. 35–36.
23. Anderson, R. J., and Anderson, M. B., Aluminium rolling-mill practice, *Chem. and Met Eng.*, vol. 22, 1920, pp. 489–491; 545–550; 599–604; 647–650; 697–702.
24. Anderson, R. J., Analysis of losses in aluminium shops, *The Foundry*, vol. 48, 1920, pp. 989–992; and *idem*, vol. 49, 1921, pp. 16–21; 54–57; 104–111; 143–147; 188–191; 235–239.
25. Pilling, N. B., and Bedworth, R. E., The oxidation of metals at high temperatures, *Jour. Inst. of Metals*, vol. 29, 1923, pp. 529–582.
26. Weil, W. M., The use of secondary aluminium in foundry practice, *Trans. Amer. Foundrymen's Assoc.*, vol. 30, 1923, pp. 613–616.
27. Anderson, R. J., Aluminium and aluminium-alloy melting furnaces, *Trans. Amer. Foundrymen's Assoc.*, vol. 30, 1923, pp. 562–604; and *The Foundry*, vol. 50, 1922, pp. 737–741; 792–799; 823–826; 866–870; 919–924.
28. Rosenhain, W., and Archbutt, S. L., The use of fluxes in the melting of aluminium and its alloys, *The Metal Ind.* (London), vol. 24, 1924, pp. 419–421; abst. of paper before The Faraday Soc.

CHAPTER XI.

ALUMINIUM-ALLOY FOUNDRY PRACTICE.

The aluminium-alloy casting industry is only about 25 years old, and its greatest growth has come in the past 15 years. Aluminium-alloy castings were made prior to 1900, but in very small quantities, and the magnitude of the industry then was insignificant as compared with its present proportions. As pointed out previously, about 50 per cent of the world's production of aluminium goes into the manufacture of light aluminium-alloy sand castings, and on the basis of the present output of aluminium, probably about 250,000,000 lb. of castings are now made annually. The automotive industry is the heaviest consumer of light alloy castings, but there is scarcely an industry or an engineering trade in which such castings are not used. Practically no substantially pure aluminium is used as such for sand castings, although a few minor applications of actual aluminium castings have been reported. In trade parlance, aluminium-alloy castings are usually referred to as " aluminium " castings, but it is as incorrect to so speak of these castings as it would be to call brass or bronze castings by the term " copper " castings.

In the United States, aluminium-alloy sand castings are made in non-ferrous jobbing foundries, in automotive foundries so-called (foundries which are departments of automobile and truck-manufacturing companies), in foundries connected with large manufacturing companies such as in the electrical industry, in vacuum-cleaner plants, and in cooking-utensil plants, as well as elsewhere in small amount. The aluminium-alloy casting industry was developed principally in brass shops, and both brass (and bronze) and aluminium-alloy castings are made in the same shops at the present time. In more recent years a number of exclusive aluminium-alloy foundries have been built, and the tendency is toward the employment of exclusive shops for casting the light alloys. Aluminium-alloy foundry practice has been most highly developed in the automotive and exclusive aluminium-alloy foundries, and modern production methods are in effect in these plants. Some of the exclusive aluminium-alloy foundries are large, with capacities ranging from 2,000,000 to 20,000,000 lb. of finished castings per annum. The number of plants in which aluminium-alloy sand castings are made in the United States is increasing at a rapid rate, and over the period 1920 to 1922 the increase was 331 plants. In Canada, the increase in the same period was 59 foundries. Some recent data gathered by *The Foundry* show that in 1922 there were 2,500 foundries and foundry departments of manufacturing plants in which aluminium-alloy sand castings were being made. Table 85 shows numerically the gain in aluminium-alloy foundries in the United States and Canada over the two-year period 1920 to 1922, and Fig. 117 is a map showing the distribution of the foundries by states. In 1922, the leading states as to number of aluminium-alloy foundries were as follows:

State.	Number of foundries.
New York	239
Pennsylvania	235
Ohio	208
Illinois	202
Michigan	172
Massachusetts	120
California	116
Wisconsin	116
New Jersey	94

In Canada there are 140 foundries in the Province of Ontario.

TABLE 85.—*Growth of aluminium-alloy foundries,[a] 1920 to 1922.*

United States and Canada, 1922	2753
United States and Canada, 1920	2363
Increase	390
United States, 1922	2500
United States, 1920	2169
Increase	331
Canada, 1922	253
Canada, 1920	194
Increase	59

[a] According to data gathered by *The Foundry*.

Since 1922 there has been a rather marked decrease in the total number of foundries in which aluminium-alloy castings are poured. Thus, in 1924, there were 2,270 plants in the United States and 206 in Canada. These figures may be compared with those given for 1920 and 1922 in Table 85.

In the United States, aluminium-alloy foundries vary in size from small shops employing one or two molders and using a small crucible furnace up to the largest concern in the business (the Aluminum Co. of America), which has four sand-casting plants and a rated capacity of 30,000,000 lb. of finished castings

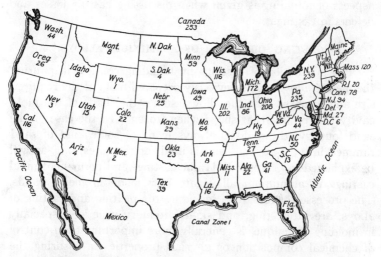

FIG. 117.—*Distribution by states of foundries and foundry departments of manufacturing plants making aluminium-alloy sand castings, 1922.*

per annum. Reverting to the question of the distribution of aluminium-alloy foundries in the United States, the map in Fig. 117 will be found useful for reference, but it should be pointed out that although New York and Pennsylvania lead in actual number of foundries in which aluminium alloys are melted, the bulk of the output comes from Michigan and Ohio, principally from the Detroit and Cleveland districts. About 50 per cent of the domestic production of aluminium-alloy sand castings comes from the Detroit district.

It is, of course, evident that it will be impossible to treat the subject of aluminium-alloy founding adequately in a single chapter, and the general details of foundry practice must be passed over with brief consideration, if touched upon at all, and only the general aspects of the subject can be dealt with here. There are many texts on founding now available which may be consulted for technical details as to foundry practice in general.

So far, little reliable technical information has appeared on the subject of aluminium-alloy founding. In the present chapter it is the object of the author to give only a broad purview of aluminium-alloy foundry practice as it is now carried out, but in Chapter XII some additional information on certain aspects of founding is given when discussing casting losses and defects in castings.

CASTING QUALITIES OF ALUMINIUM ALLOYS.

The production of castings by founding consists in pouring the melted alloy into a previously prepared mold of the required size and form, so that the alloy assumes the shape of the mold cavity and retains that shape when frozen. In the case of the sand foundry, the mold is always made of some kind of sand rammed into a box, and sand is used for the cores, but, as will be explained in Chapter XIII, die castings and permanent-mold castings in aluminium alloys are now made in metal molds. The processes of founding, including the melting and casting of alloys, are interesting, and skillful foundry practice, especially in molding technique, is generally more important than control of chemical composition or physical properties in ensuring the production of satisfactory castings with a minimum of wasters. Before dealing with the technique of aluminium-alloy foundry practice, it is of importance to discuss briefly the casting qualities of aluminium alloys. From the point of view of founding it is of great importance to select alloys which possess good casting properties,[52] and it is of interest here to mention briefly some of the factors which determine the casting qualities of any alloy.

Turner,[43] in his interesting paper on the casting of metals, has discussed a number of factors which affect the casting properties of alloys. In considering the technique of founding and the difficulties encountered in the production of aluminium-alloy sand castings, it is clearly advisable to appreciate the variations that may exist in the casting qualities of the various alloys and the factors affecting them.

In order to obtain sharp castings, i.e., good impressions of the molds, and to fill up all parts of the mold cavity, it is necessary that the alloy should be well fluid when liquid, and it should flow freely and rapidly. This pre-supposes high fluidity and low viscosity in the alloy, but comparative values for these properties in the case of alloys are scant. Measurements of the fluidity and viscosity of the light aluminium alloys, and the determination of the effect of added elements on these properties would be found useful in the development and selection of alloys with good casting properties. Fawsitt [7] and others have studied the viscosity of a few metals, finding that the viscosity is small, being only a few times that of water, and that it diminishes with rising temperature. Thus, the viscosity of water at 12° C. is 0.128, while that of lead at 347° C. is 0.0293. It follows that for all practical purposes there is not much difference between the rates of flow of equal weights of water and of liquid metals.

The reason why water wets a sand mold while liquid alloys do not is explained by the forces of surface tension and cohesion. Considerable work has been performed on the surface tension of liquid metals, but this has not yet been extended to alloys, and determinations in respect to the light alloys would serve a useful purpose. It may be added that there is a special surface effect in the case of aluminium-rich and aluminium-containing alloys, viz., the formation of a surface skin owing to oxidation; this skin may be sufficiently strong and tough so as to prevent the flow of the alloy into thin parts and sharp angles of a mold on casting. A thin tough adherent film of oxides forms on the surface of some light aluminium alloys on melting, particularly at high temperatures in the case of nickel-containing aluminium alloys, and this film may ruin the alloy for casting purposes. In the case of heavy aluminium alloys, e.g., 10 per cent aluminium bronze, a tough film of aluminium oxide forms on the surface, and this gives rise to much difficulty in casting. This question has been investigated by Carpenter and Edwards,[8] and it is of interest to consider their view of the matter. They point out that aluminium bronze should be poured as quietly as possible, and that care should be taken to prevent agitation of the alloy after it enters the mold. The occlusion of dross particles in aluminium-bronze castings is explained by these investigators as follows: If the alloy enters the mold quickly, ripples or waves are formed which break through the skin and expose fresh surfaces of alloy which are immediately oxidized. The continual recurrence of this process causes the formation of a large amount of dross. The skin of aluminium oxide formed on this alloy is tenacious and refractory, and consequently when two or more ripples overlap, the skin prevents the liquid alloy from uniting. This explains incidentally why a large amount of dross is formed when liquid light aluminium alloys and aluminium bronze are stirred.

It should be added that in connection with sand difficulties and the selection of sands for founding different alloys, the question of the searching properties of alloys is important. The burning-on of steel has been encountered to a considerable extent in steel-casting practice, and has been eliminated by the selection of suitable sands. Some alloys, notably lead-rich alloys, have high searching qualities, and readily penetrate into small veins and pores in a sand mold. High searching values correspond to low surface tension; water has high searching qualities, and alloys with low surface tension readily wet the sand. Searching and penetration of sand by very mobile alloys can be prevented by employing sand of small grain size and close bond, and by hard ramming. Of course the natural venting qualities and permeability of a mold to the escape of gases are sacrificed by employing close sand and by hard ramming. The surface tension of liquid aluminium is 520 dynes per cm., as compared with 1,178 dynes for liquid copper and 73.5 dynes per cm. for water (at 15° C.). The latent heat of fusion is an important factor in determining the casting qualities of an alloy, and other conditions being the same, the time period required for the solidification of a metal depends upon its latent heat of fusion. The latent heat of fusion of aluminium is the highest of any of the common metals, and this explains in part why aluminium and its light alloys remain pasty for a long time on cooling in the freezing range. Part of the pastiness is caused by oxidation. The question of thermal expansion of the alloys is important, since this property is the reverse of the solid contraction (cf. Chapter IV). The large solid contraction of the light aluminium alloys causes some of the principal difficulties in foundry practice, giving rise to cracked castings, as well as warped, distorted, and off-size castings. The production of satisfactory castings in aluminium alloys, with a minimum of wasters, depends upon the thorough recognition of the following three factors: (1) the high contraction in volume; (2) the pasty stage that exists near the melting point, and which persists for a relatively long time; and (3) the low strength at elevated temperatures. It may be said that it is possible to increase the strength of given alloys at high temperatures by the addition of small quantities of certain metals. Thus, additions of 1 to 2 per cent of either manganese, nickel, or iron increase the strength of some light aluminium-copper alloys at elevated temperatures, which fact is taken advantage of in foundry practice.

In founding, the usual device employed to secure dense castings is the provision of a casting head of sufficient height and cross-section, the general idea being that a large casting head exerts considerable pressure on the casting and thus leads to increased density; however, the pressure exerted by an ordinary riser is negligible, and its effect is simply to feed liquid alloy to the casting, thus minimizing the effect of the liquid shrinkage. A riser of an aluminium alloy 2 ft. high would exert a pressure of only about 3 lb. per sq. in., so that the riser pressure exerted is of no account in influencing the soundness of castings. Of course, this small pressure might be sufficient to prevent the escape of gas in the freezing alloy, thus eliminating blowholes. The question of the effect of pouring temperature upon the contraction of aluminium alloys and upon their strength is important, and for aluminium alloys, like other alloys, there is a range of pouring temperature which yields the most satisfactory results both from the point of view of casting and physical properties. Apparently, an intermediate pouring temperature is more desirable than a high or too low temperature, considering all aspects of the problem. The effect of pouring temperature upon the strength of aluminium alloys has been discussed by Gillett;[11] the effect of the pouring temperature upon soundness in the resultant castings has been dealt with by the author;[30] and the effect of the pouring temperature upon the contraction of aluminium alloys has also been discussed by the author.[52] The various aspects of the question can not be discussed in detail here, but the pouring temperature should be kept low—in the range 600 to 800° C., depending upon the alloy and the type of casting. Hence, the burning-on difficulties which occur in steel practice are not known in light alloy founding. The pouring temperature is of interest in relation to the thermal conductivity and thermal capacity of molds. Metals of high melting point dissipate heat more quickly than those of lower melting point when poured into a mold. Ordinarily, there is no danger that aluminium alloys, when poured at the ordinary temperature, will exhaust the thermal capacity of a mold before final solidification, although their specific heat is high. The pouring temperature is determined by the melting point of the alloy, among other factors, and by its viscosity at different temperatures. Thus, high iron in aluminium-copper alloys raises the melting point considerably and apparently greatly increases the viscosity, so that such alloys give much difficulty when cast at the ordinary temperature employed for light aluminium-copper alloys.

As pointed out in Chapter IX, when discussing gas occlusion by aluminium and its alloys, the problem of the dissolution of gases by metals and alloys is one that is little understood, particularly as regards the non-ferrous class of materials. The fact that castings frequently contain blow holes has led, of course, to the study of the relationships which exist between alloys and gases under various conditions. The light aluminium alloys are normally very unsound, and the reasons for their relatively great unsoundness, as compared with brass and bronze, are not understood. In development work in the light alloys it would serve a useful purpose if the relative soundness of a series of the alloys on casting were studied. The question of the soundness of alloys is closely connected with their real density, as distinguished from their true theoretical density, and this consideration is important in the selection of aluminium alloys for leak-proof parts and for other parts which must be non-porous. Turner has given much attention to the question of the density of alloys, pointing out that when there is no freezing interval, the density is the same whether the metal or alloy is slowly or quickly cooled; but that where there is a freezing interval, i.e., where there is a temperature range between the liquidus and solidus, sand-cast bars give a lower density than chill-cast ones. Thus, slower cooling gives lower density than rapid cooling, and sand-cast bars show expansions when examined by the extensometer. Little is known as to these effects in the case of the light aluminium alloys, and study of the question is warranted. While the casting properties of aluminium alloys can not be discussed satisfactorily on the basis of the available information, it may be stated broadly that the most suitable alloy for foundry work, other properties being relatively the same, is the one having the least contraction on freezing.

SOME PRACTICAL ASPECTS OF FOUNDING.

As indicated, the more detailed and practical aspects of castings production can not be treated in the present book, but it is of importance and interest to discuss briefly some of the more important items that bear on aluminium-alloy founding. The items that appear for consideration are the following, and these are treated *in seriatim* below: (1) kinds of castings made; (2) kinds of alloys used; (3) composition of melting charges; (4) melting methods; (5) fluxes employed; (6) effect of foundry practice on the quality of castings; (7) pyrometry; and (8) defects in castings.

Kinds of Castings Made.—In discussing the uses and applications of aluminium alloys in Chapter VII, most of the different kinds of castings produced in sand practice have been mentioned or alluded to, and this subject need be discussed only briefly here. While the automotive industry was the important factor in promoting the employment of aluminium-alloy castings in machine and engine design, the World War accelerated this to a large degree. During that period aluminium alloys were applied in the construction of trucks, motor buses, cars of all kinds, aircraft (both sea and land), and oil-burning engines for submarines and destroyers. A great variety of sand castings is now made, and the bulk of the production goes into automotive construction in the manufacture of internal-combustion engines for passenger cars, motor trucks, and aircraft. Sand-cast aluminium alloys are used generally in the engineering trades as substitutes for brass, bronze, gray-iron, and steel castings, and the principal applications, outside of automotive construction and aircraft, are in electrical manufacturing, household appliances, vacuum cleaners, cooking utensils, and in the chemical industry. The many specific kinds of aluminium-alloy castings made are too numerous to be cited here (cf. Chapter VII).

Table 86 shows the total amounts and the average percentage of the total of various kinds of aluminium-alloy sand castings made in the United States in 1919. Of course, the annual amount of aluminium used for various different kinds of castings differs from year to year, depending upon industrial and other conditions. In 1918, practically the entire production of castings was for aircraft and motor cars, while in 1919, the amount going into motor cars was exceptionally large, because of the prosperity in the motor industry and the continued tendency toward the use of aluminium alloys in automotive work.

TABLE 86.—*Total output of various kinds of aluminium-alloy sand castings in 1919.*

Kind of castings.	Output, lb.[a]	Per cent of total output.
Automotive [b]	75,000,000	92.59
Vacuum-cleaner	3,500,000	4.32
Cooking-utensil	1,480,000	1.83
Small electric motor housings and cases	250,000	0.31
Patterns and flasks	200,000	0.25
Scale trimmings	100,000	0.12
Machinery parts	50,000	0.06
Ornamental	40,000	0.05
Scientific and other instruments	30,000	0.04
Miscellaneous, not classified	350,000	0.43
Total	81,000,000	100.00

[a] On the basis of figures reported to the U. S. Bureau of Mines, and from other sources.
[b] Including all kinds of castings for the motor industry—crankcases, oil pans, camshaft housings, manifolds, carburetor bodies, carriers, brackets, etc.

According to data gathered by the author [40] for the U. S. Bureau of Mines, about 81,000,000 lb. of finished aluminium-alloy sand castings were made during 1919 in the United States, and the consumption is now at the rate of 100,000,000 lb. annually. The estimated capacity of the foundries in the United States is about 150,000,000 lb. (1924 basis). Castings made in aluminium alloys vary in weight from very small parts weighing less than 1 oz. up to parts weighing 1,500 lb. or more. According to statistical data reported to the U. S. Bureau of Mines in 1919, the weight of the lightest sand casting reported was 0.017 lb. (0.272 oz.), and many kinds of small castings weighing 0.25, 0.5, 1, and 2 oz. are made. The weight of the heaviest casting reported was 600 lb., but very large castings weighing several thousand pounds are planned for production by the General Electric Co. Crankcases weigh from 40 to 100 lb., depending upon the size and design.

Kinds of Alloys Employed.—As indicated in Chapters VI and VII, many kinds of light aluminium alloys are used for sand castings, but the standard No. 12 alloy, of the nominal composition 92 : 8 aluminium-copper, is the one usually employed. On the basis of the figures reported to the U. S. Bureau of Mines, out of a total annual output of 81,000,000 lb. of sand castings, 78,897,000 lb. were made of No. 12 alloy, or 97.4 per cent. Numerous other alloys are used in the United States for sand castings,[40] but the total output is relatively small. The total output of castings made of various aluminium alloys in 1919, as reported to the U. S. Bureau of Mines and from other sources is shown in Table 87.

As indicated elsewhere in this book, various aluminium-copper alloys containing 2 to 14 per cent copper are employed for sand castings, and the binary aluminium-copper alloys as a class are the most important. Binary aluminium-magnesium alloys are used in small amount, but binary aluminium-silicon alloys have come into fairly wide use during the past two years. Binary aluminium-zinc alloys are not used to any extent in the United States now, although they were formerly prominent alloys. They are still favored in England for many purposes, but they are gradually being supplanted everywhere by the light aluminium-copper alloys. Ternary aluminium-copper-zinc alloys, of various compositions, occupy the second position in total output; certain of these alloys are desirable for particular purposes, such as carburetor bodies. Ternary aluminium-copper-manganese alloys, e.g., 97 : 2 : 1 aluminium-copper-manganese, are employed slightly for sand castings. Ternary aluminium-copper-silicon alloys have been developed recently, and these are now used to a subordinate extent for sand castings. These alloys possess casting properties superior to any light alloy, with the exception of the binary aluminium-silicon alloys. Some alloys of aluminium, copper, and tin are employed, particularly for castings which must be highly polished, and aluminium-copper-nickel alloys are used to a small extent for sand casting. Conditions in the casting industry have changed considerably

TABLE 87.—*Kinds of alloys melted for sand castings and percentage of total output of each, in 1919.*[a]

Nominal composition.	Total output, lb.	Percentage of total output.
Binary aluminium-copper alloys		
92 : 8 Al-Cu alloy	78,897,000 [b]	97.403
98 : 2 Al-Cu alloy	4,000	0.005
97 : 3 Al-Cu alloy	1,000	0.001
96 : 4 Al-Cu alloy	225,000	0.278
94 : 6 Al-Cu alloy	40,000	0.049
93 : 7 Al-Cu alloy	200,000 [b]	0.247
90 : 10 Al-Cu alloy	200,000 [b]	0.247
Binary aluminium-magnesium alloys		
92 : 8 Al-Mg alloy	25,000	0.031
Binary aluminium-zinc alloys		
90 : 10 Al-Zn alloy	30,000	0.037
74 : 26 Al-Zn alloy	20,000	0.025
67 : 33 Al-Zn alloy	100,000 [c]	0.123
Ternary aluminium-copper-manganese alloys		
98 : 1 : 1 Al-Cu-Mn alloy	15,000	0.019
Ternary aluminium-copper-tin alloys		
88 : 8 : 4 Al-Cu-Sn alloy	25,000	0.031
85 : 5 : 10 Al-Cu-Sn alloy	12,000	0.015
Ternary aluminium-copper-zinc alloys		
90 : 5 : 5 Al-Cu-Zn alloy	20,000	0.025
90 : 8 : 2 Al-Cu-Zn alloy	50,000	0.062
90 : 7 : 3 Al-Cu-Zn alloy	3,000	0.004
87 : 8 : 5 Al-Cu-Zn alloy	5,000	0.006
80 : 7 : 13 Al-Cu-Zn alloy	2,000	0.003
78 : 3 : 19 Al-Cu-Zn alloy	1,000,000 [d]	1.234
76 : 4 : 20 Al-Cu-Zn alloy	5,000 [d]	0.006
73 : 2 : 25 Al-Cu-Zn alloy	1,000	0.001
65 : 3 : 32 Al-Cu-Zn alloy	70,000	0.086
Ternary aluminium-manganese-zinc alloys		
62 : 1 : 37 Al-Mn-Zn alloy	50,000	0.062
Totals	81,000,000	100.000

[a] On the basis of data reported to the U. S. Bureau of Mines and from other sources.

[b] The figures for No. 12 alloy, for 93 : 7 aluminium-copper alloy, and for 90 : 10 aluminium-copper alloy may be subject to some correction because of errors in reporting. Thus, the figures for the 93 : 7 aluminium-copper alloy might be included with those for No. 12 alloy, and the figures for No. 12 alloy may include output of 93 : 7 aluminium-copper, 90 : 10 aluminium-copper, and other alloys richer in copper, such as manifold and piston alloys.

[c] Figures for binary aluminium-zinc alloys vary markedly from year to year.

[d] Figures for ternary aluminium-copper-zinc alloys vary markedly from year to year depending upon the demands of the trade.

NOTE: This table is as accurate as the data available permit.

since Table 87 was constructed and at the present time (1924) about 50 per cent of the aluminium-alloy castings made in the United States are poured in the 92 : 8 aluminium-copper alloy, about 40 per cent in 90 : 7 : 1 : 2 aluminium-copper-iron-zinc alloy, and the remaining 10 per cent in other alloys. The 95 : 5 aluminium-silicon alloy is gaining ground as a general casting alloy, and ternary aluminium-copper-silicon alloys containing 3 to 5 per cent each of copper and silicon are being used to some extent. So-called "Y" alloy (92.5 : 4 : 1.5 : 2 aluminium-copper-magnesium-nickel) is being used slightly. An insignificant amount of castings are made in complex alloys.

Composition of Melting Charges.—The composition of the melting charge in foundry practice depends, of course, upon the alloy melted, but even in casting No. 12 alloy the practice in different foundries varies widely. In some plants only primary aluminium pig (plus the necessary intermediate alloy or copper) is used together with a small amount of foundry scrap made up of gates, risers, and defective castings originating in the plant; in others, part secondary aluminium-alloy pig is used with the preceding charge. In some plants primary or secondary No. 12-alloy pig is employed mainly, while in others scrap castings purchased from outside sources are used whenever possible. So far as there is any evidence of standard practice, in numerous large and small foundries a mixed charge is made up of the following materials: primary aluminium pig, secondary aluminium-alloy (No. 12 alloy) pig, 50 : 50 copper-aluminium alloy, and foundry scrap originating in the plant and consisting of gates, sprues, risers, and defective castings.

The average percentage of materials charged per heat varies, as stated, but the figures in Table 88 show the tendencies in making up heats of No. 12 alloy. In the average foundry it is imperative to re-melt all the foundry scrap originating in the production of castings, and heavy foundry scrap (gates, risers, and defective castings) is a highly desirable material for melting. One foundryman claims to be able to " tone up " inferior aluminium pig by a proper admixture of No. 12-alloy scrap in the charge; and some founders prefer to make up heats from scrap castings exclusively, provided they can obtain scrap of high quality, such as scrap airplane-motor castings. In the majority of the foundries having large output, from 20 to 80 per cent of

TABLE 88.—*Tendencies in the composition of charges in making up heats of No. 12 alloy.*

Quantity per average heat, lb.	Average percentage of materials charged per heat.[a]							
	Primary aluminium pig.	Copper.	50 : 50 Cu-Al alloy.	Secondary aluminium pig.	Primary No. 12-alloy pig.	Secondary No. 12-alloy pig.	Foundry scrap (gates, risers, and defective castings).	Scrap castings from outside sources.
4000	33.6		6.4			25	35	
600	60		9.6			16	14.4	
400						50	50	
250	8		8.5	50		3.5	30	
200	50		8				30	12
200						40	35	25
150					45		20	80
120	33.6		6.4			20	20	
50							(?)	
50	92	8					(?)	
40						70	30	
40					100		(?)	
40				92			(?)	100
35	30		9.6	30			30.4	
35		2.5		30		52.5	15	
30	44		(?)			44	12	

[a] Reported as being used in various foundries.

primary aluminium pig, plus other materials, is used, but some founders prefer to use primary No. 12-alloy pig rather than to make the alloy in the foundry. Moreover, it should be said that the condition of the primary and secondary metal markets will affect the kind of materials used in heats; purchases of melting stock may be made in anticipation of rising or falling markets under normal conditions, or of materials that may be temporarily obtained cheaply, and the foundry foreman is required to melt the material supplied. In plants where machine shops are operated for machining aluminium-alloy castings, in addition to the foundry department, borings may be sent out on toll to a smelter and run into pigs; in such plants varying amounts of secondary No. 12-alloy pig are employed in the charges.

Unfortunately, too little attention is paid in the average foundry to the make-up of melting charges for sand-casting practice, although this matter is one of the utmost importance and deserves the most careful attention in actual production. Too often it is dismissed on the ground that the preparation of an alloy is merely a melting operation which is too simple to require anything but the most casual attention. The facts in the case are that the metallurgical principles involved in alloying may be made the basis for much more efficient practice. The question, as applied to the preparation of light aluminium-copper alloys, has been discussed at length in Chapter VIII. The necessity for charging *clean* scrap can not be urged too strongly. Too often, little or no attention is paid to this matter, and heats are often made into which chunks of core sand, pieces of brick, nails, core wires, chills, dirt, and other foreign materials are charged. It is too much to expect that sound and strong castings can be poured from dirty heats. All foundry scrap and defective castings that are returned to the melting room should be thoroughly freed from sand; and light pieces of scrap, such as splashings and spatters, should be charged by fork so that fine dirt will drop through the tines. Aluminium alloys must be handled cleanly in melting to ensure that the resultant castings will be as free as possible from hard spots and inclusions (cf. Chapter XII). Dirty charges may be responsible also for porosity and leaks. It is advisable to make chemical analysis of all melting stock, and the control of composition of successive melts can be made by analysis of periodic samples. Also, drillings from gates taken from castings, according to a systematic sampling procedure, may be analyzed. It has been pointed out by the author [31] that in spite of the fact that primary aluminium pig may contain over 99 per cent aluminium (on the basis of the difference method of analysis), the mechanical properties and behavior in the foundry may be very variable (cf. Chapter III). Where physical specifications are to be met, bars must be tested for tensile strength, and standard test bars should be used for control tensile tests.

Melting Methods in Foundry Practice.—Melting methods in foundry practice, and the types of furnaces employed, have been discussed at length in Chapter IX. While these subjects do not require further treatment here, it is of interest to discuss briefly the question of melting ratio, since the melting ratio in aluminium-alloy founding is normally rather high. As is known, castings made in aluminium alloys are normally gated with wider gates, and larger, heavier, and higher risers are employed than is the case in ordinary brass and bronze founding. Consequently, the melting ratio in aluminium-alloy work may be expected to be higher than in brass and bronze practice. In numerous foundries, from 1.5 to 2 lb. of an aluminium alloy is melted for every pound of finished casting produced. This means, really, that from 0.5 to 1 lb. of alloy appears in the form of sprues, runners, risers, and defective castings, and this figure also includes the melting loss due to oxidation and shrinkage. Taking the melting loss as 1.5 per cent, and the defective casting loss as normal, the indicated average melting ratio for 50 foundries in the United States was found to be 1.6, which means that for every pound of finished casting produced, 1.6 lb. of alloy was melted. Reported figures for the melting ratio of 50 foundries varied between 1.12 and 2.21 per cent.

Fluxes Used in Foundry Melting.—Many fluxes have been used in melting aluminium alloys for the purpose of cleaning the bath and freeing it from entangled foreign matter (principally dross). Zinc chloride is still used to a considerable extent for these purposes, and while it apparently has some value, the tendency in recent years in the United States has been toward discontinuing the use of this or any other flux. Zinc chloride is normally used in amount from 0.02 up to 0.25 per cent by weight of the charge, and it is added after the melt is liquid and about ready to be poured. Ammonium chloride (sal-ammoniac) is also used to a small extent, as are cryolite, sodium chloride, fluorspar, and borax. Potassium nitrate has been employed, and many mixtures of salts, e.g., 88 : 12 sodium chloride-sodium fluoride and 60 : 40 potassium chloride-cryolite, have been used. As explained in Chapter VIII, when discussing the preparation of magnesium-bearing aluminium alloys, magnesium is used considerably in melting practice, being added with the object in view of deoxidizing the bath. This is quite useless, since magnesium will not rob aluminium oxide of oxygen except under special conditions and at high temperature, and the heats of formation of magnesium oxide and aluminium oxide lie close together. The heat of formation of aluminium oxide is ordinarily taken as 128,700 cals., calculated on a gm.-atom basis, as compared with 146,000 cals. for magnesium oxide; although Biltz and Hohorst give the heats of formation as 113,100 cals. for magnesium oxide and 125,000 cals. for aluminium oxide. However, magnesium will reduce both carbon monoxide and carbon dioxide at normal melting temperatures, and consequently serves as a degasifier. A small addition, i.e., 0.25 to 0.5 per cent, is sufficient to degasify an ordinary melt, and the addition of magnesium also yields finer grain size in the resultant alloy and increased strength.

Effect of Foundry Practice on Quality of Castings.—The question of defects in aluminium-alloy castings is discussed in Chapter XII, and here it is of interest to discuss briefly the methods of obtaining good sound castings of satisfactory mechanical properties. In general, the principles applied to maintain uniformity in tensile-test bars of good mechanical properties can be applied with equal advantage to castings which are complicated, and the ideal conditions under which test bars are made should be aimed at so far as is possible in all castings made to specifications. The melting stock should be selected with as full data as to its previous history as it is possible to obtain, and this should be selected and mixed so as to give the desired chemical composition. All melting stock should be under strict chemical control, and any stock which is of poor composition or which has been overheated, or otherwise mistreated, should not be employed. The alloy should be melted in a non-oxidizing or neutral atmosphere, care being taken that no part of the charge be overheated. The melt should be raised only to the minimum temperature required for the necessary handling, so that it is brought to the molds at, or slightly above, the correct pouring temperature. Aluminium alloys should be melted rapidly and poured as soon as possible after attaining the correct melting temperature. Charges should not be allowed to " soak " or stand in the furnace in the liquid state for long periods. Molds should always be prepared and ready, waiting to be poured, rather than to have metal standing in the furnace waiting for the molds. The alloy should be cast at the proper pouring temperature, in accordance with the principles given below under Pouring Temperatures and Pyrometry. Since the above precautions are taken in handling the alloy prior to pouring, it is obvious that the mold into which the alloy is poured should be carefully prepared; one of the most important considerations is that of cleanliness. Among other factors, the following are of the utmost importance as regards the preparation of molds: (1) the gates should be properly proportioned, correctly distributed, and placed in the right position; (2) the gates should be designed so as to choke and skim the alloy before it enters the mold cavity; and (3) there should be a sufficient number of risers, properly placed, and of the correct size to feed all sections.

Blowholes, porosity, and unsoundness in castings and poor mechanical properties occur together, and the general principles of founding applied to produce good mechanical properties will tend largely to the elimination of porous and leaky castings. While the use of fluxes, so-called, in aluminium-alloy melting is decreasing, it appears that the addition of a small amount of zinc chloride to a heat, followed by stirring and skimming, is of definite advantage. Tests of various kinds, including chemical, physical, and metallurgical, are of value in foundry operations for several reasons, viz., (1) to control the composition of the alloys; (2) to control the melting methods; (3) to check the quality of the molding practice; and (4) to provide data for design.

Pouring Temperatures and Pyrometry.—As pointed out in earlier chapters, the light aluminium alloys are very susceptible to pouring temperature as regards strength, soundness, and general properties, and the general rule in foundry practice is that all castings should be poured at as low a temperature as is consistent with the alloy filling the mold and avoiding cold shuts. In actual practice, the amount of pyrometric control in different foundries varies from not any to very careful control, where temperature measurements are taken of both the melting charge and the alloy in the pouring ladles, i.e., in the best practice both melting and pouring temperatures are controlled as contrasted with the worst practice where pyrometers are not used. In some foundries, only the temperature of the liquid alloy in the furnace is taken when the charge seems sufficiently hot, while in others no attention is paid to this, but the actual pouring temperature at the molds is controlled. In the best practice for controlling melting temperatures, the temperature of the charge is taken at frequent intervals to ensure keeping the temperature below some prescribed maximum. Close pyrometric control during both melting and pouring is a necessity in maintaining quality, and this means the elimination of at least two of the many variables in foundry practice.

As just mentioned, aluminium-alloy castings should be poured at as low a temperature as will permit the alloy to fill the mold cavity readily without danger of cold shuts. Taken by and large, a casting poured at a low temperature, so far as it is affected by the pouring temperature, will be the most sound (as shown by water and gasoline porosity tests), the strongest, and most ductile that can be produced. While cooling to the correct pouring temperature from a high melting temperature does much to counteract the harmful effects of overheating, it has been proved experimentally that castings poured from heats melted at a low temperature are invariably better than those poured from heats melted at a high temperature and then cooled to the correct casting temperature.[30] The observation that castings poured at a low temperature are better as to physical properties and soundness than those poured at a high temperature may be explained usually by the fact that the casting poured at a low temperature is fed better, especially in the heavy sections, than one poured at a high temperature. Liquid metal is lower in density the higher the temperature, i.e., it expands with rising temperature and contracts with falling temperature. If an alloy is poured into a mold cavity at a high temperature, the contraction of the liquid alloy as it cools to the freezing point, which must be fed up, is greater than the contraction from a lower temperature. A greater weight of alloy can be poured into a mold cavity of a given volume at a low temperature than at a high temperature, and this means, for example, that the risers suitable for a casting poured at 700° C. may not be sufficiently large to feed up, i.e., take care of the shrinkage, in the same casting poured at 800° C. As will be shown later when discussing the question of gating, so far as is feasible and consistent with other conditions, molds should be gated in such a way as to permit using a low pouring temperature. By so doing, the risk of defects such as cold shuts, porosity, and draws occurring will be materially reduced.

While a low pouring temperature is the *desideratum* in light alloy founding, certain general exceptions to this rule may be explained. Thus, for example, certain castings, which are molded so that they have a large surface in the cope to be machined, will yield better surfaces on machining and have most desirable machining qualities when poured at a temperature somewhat higher than that which is necessary to just pour the castings. At the same time, this condition can be circumvented, avoiding a higher pouring temperature, by regulation of the speed of pouring and by gating so as to avoid violent agitation of the alloy in the molds, which yields a collection of dross and oxide scum in the cope. When the sand is cold, for instance, on winter mornings, it is necessary to use a higher temperature than under warm weather conditions. Very thin castings must be run at higher temperatures than thick castings owing to the chilling effect of the sand, and in all cases the pouring temperature for a given casting must be determined empirically. An intermediate pouring temperature yields lower contraction than a high or low pouring temperature, so that the correct pouring temperature for any given casting can be determined only after pouring a number of castings at different temperatures and comparing the results. Gillett[11] has dealt with the effect of pouring temperature upon the strength of various aluminium alloys, and has shown that practically all of the light alloys are extremely susceptible to this factor. A general dictum is the higher the pouring temperature the lower the strength and elongation, and there may be so much as 5,000 lb. or more difference in test bars poured at high and low temperatures—say 200° C. apart.

Aluminium-Alloy Foundry Practice 501

Liquid aluminium alloys are very destructive to most types of pyrometers, and so far much difficulty has been experienced in obtaining actually satisfactory protecting tubes. Various forms of thermocouples are in use in foundry practice. A bare base-metal (nickel-chromium) thermocouple inserted into the alloy is employed to some extent. The couple is of the forked type, and this gives good results, since the fork prevents a bridge of the alloy being formed across the two elements. The pyrome-

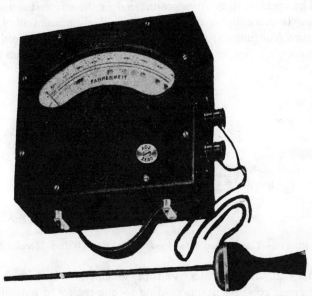

FIG. 118.—*Pyod and portable indicator for aluminium alloys (Wilson-Maeulen Co.).*

ter first used by Gillett in work at the former Aluminum Castings Co., and which has been used extensively in many foundries, is the pyod (Wilson-Maeulen). One element of the pyod is the outside tube of the pyrometer, while the other element is an internal wire on which is wound asbestos insulation. This construction makes the thermocouple more sensitive to temperature changes than the ordinary wire couple with a protective tube, since the protective tube causes a lag in the temperature record. Pyods are intended for continuous service up to 900° C. (1,650° F.), or for intermittent service up to 1,000° C. (1,800° F.). Pyods have been employed by the author for a number of years, and are regarded as the most suitable pyrometer for use in aluminium-alloy foundry practice. Fig. 118 shows a pyod and portable indicator, while Fig. 119 shows the construction of the pyod. In Fig. 119, B is the center wire element, D is the asbestos insulation, P is the outside tubular element, and W is the welded end joining the two elements. The Thwing pyrometer (type A5X) is also used in aluminium-alloy work, as well as other makes.

The greatest difficulty encountered in the successful application of thermocouples to aluminium and aluminium-alloy temperature (continuous) measurements, when employing ordinary wire· couples, lies in obtaining a satisfactory protection tube.

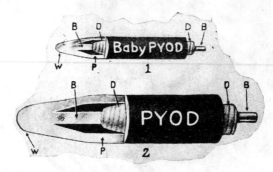

FIG. 119.—*Construction of the pyod; actual sizes (Wilson-Maeulen Co.).*

Usalite (Stupakoff) protective tubes shielded by bauxite sheaths have given good results, according to one maker of pyrometers. Another maker recommends graphite as a sheath for various kinds of protection tubes. A lime wash for pyods has been employed successfully in rolling-mill work. Various types of industrial pyrometers have been described in the U. S. Bureau of Standards publication [42] on pyrometric practice.

MOLDING METHODS IN FOUNDRY PRACTICE.

Molding methods in the founding of aluminium alloys are almost as many and varied as the number of foundries, due chiefly to the personal element. In the past, and owing to the early difficulties experienced, numerous rules were laid down for the molding and casting of aluminium alloys, many of which were exceedingly contradictory. Molding methods will be taken up in only a general way here, since a thorough discussion is without the scope of this book; moreover, it is the author's intention, largely, to afford an insight into the fundamental principles involved in aluminium-alloy work rather than to deal with the production of particular kinds of castings by different molding methods. Hence, it may be stated broadly that there are certain inherent difficulties in the founding of aluminium alloys which must be surmounted for successful production, and the experience gained by foundrymen during the past twenty years indicates that aluminium alloys may be handled almost as easily as iron and steel or brass and bronze. The molding equipment used in the aluminium-alloy foundries of the United States varies from crude board hand boxes and flasks, in which molding is done on the bench or floor, to special molding machines of large capacity using aluminium-alloy mold boxes and flasks. Molding equipment for the production of cores varies similarly. Machine molding has made rapid strides during the past five years in the aluminium-alloy foundry, owing mainly to the large production per unit that can be secured. The general superiority of machine molding in the foundry, whether on large or small work, is widely recognized,[34] and no labored argument is necessary to demonstrate its worth. Light cores are still made mainly by hand, but heavy cores, such as crankcase and other body cores, are made on standard roll-over core machines in the large foundries. In crankcase molding in a modern foundry,[36] the sand, handled by suitable mechanical equipment, is delivered to hoppers over the molding machines, from which it is delivered to the mold box through a clam-shell or other gate. This method of filling mold boxes prior to ramming gives greatly increased production over the old shoveling method. The molds are handled with hand-operated cranes. Fig. 120 shows the lay-out of this foundry. A complete system of conveyors and other equipment is employed for handling and preparing molding and core sands, and for carrying rough castings from the foundry floor to the cleaning room. The overhead conveying system in operation in the plant of the Wilson Foundry and Machine Co., Pontiac, Mich., has been described.[57]

Design and Pattern Making.—The first item to be considered in the production of castings is the design of the castings and the preparation of the patterns. While in general it is practice to allow 0.156 in. per ft. for the pattern-maker's shrinkage on aluminium alloys, this is an unsafe procedure, and the actual contraction of the separate alloys should be determined and known before the patterns are designed. Some figures for the linear contraction of a series of light aluminium alloys have been given in Chapter IV, and these will be found useful in pattern design. The effect of poor design upon the resultant casting losses is discussed at length in Chapter XII. Patterns can be made from wood, brass, or aluminium alloys, and the latter are used exten-

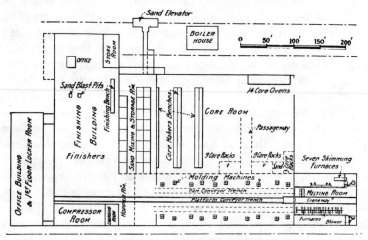

Fig. 120.—Layout of large aluminium-alloy foundry (The Foundry).

sively. The contraction in volume of most of the light aluminium alloys is large, and sufficient allowance should be made to take care of this in both the patterns and the core boxes. The pattern shrinkage allowed for aluminium alloys varies from $\frac{3}{32}$ to $\frac{3}{16}$ in. per ft.

In pattern making, as is well known, current ordinary practice usually entails making first a master pattern in wood for any desired casting, and then from this the standard metal pattern, to be employed in production, is cast. This refers, of course, to repetitive jobs where the number of castings to be made is large, and where specifications as to dimensions are close. Now, a pattern-maker, not of broad experience in aluminium-alloy work, but required to employ an approximate figure for the contraction of the alloys or to rely on data given in engineering handbooks, could not possibly make on one trial a wooden master pattern that would give cast metal patterns which would yield in turn castings of the required dimensions. Invariably, it is necessary, even in the case of skilled pattern-makers of much experience with the alloys in question, to file, scrape, or build up the metal pattern at various positions so that castings of the required dimensions can be poured. Thus, ordinarily, in the preparation of metal patterns for important repetitive work in founding, e.g., crankcases, it is necessary to proceed about as follows: First, a wooden pattern will be made and then a metal pattern cast therefrom. Then a few castings will be poured from the metal pattern to determine where the dimensions are incorrect and where the defects occur, and the required changes will be made on the wooden pattern. Then, another metal pattern will be poured from the corrected wooden pattern, and more castings poured, and so on. Corrections will continue to be made until a wooden master pattern is obtained that will give a metal pattern of the correct dimensions which, in turn, will yield castings of the size and tolerances required. In all cases, the construction of a master pattern is based on the principle of cut-and-try, and when an inaccurate value is employed for the contraction, the procedure described above may be lengthened almost interminably.

The procedure described is expensive and unscientific, and after being carried out often yields patterns which are almost certain to give rise to defective castings. Where a pattern-shop, in charge of a competent pattern-maker, is attached to an aluminium-alloy foundry, it is desirable that data should be available for the contraction of all alloys employed, and such data should include figures for the contraction of the alloys as affected by the important factors which influence the values. While even with rule-of-thumb methods an intimate knowledge of the behavior of various alloys on casting and experience with a variety of designs will permit practical pattern-makers to construct good patterns, still these methods are technically incorrect. As already indicated, the view which it is desired to put forward here is that data for the construction of the various alloys should be of especial value in pattern practice, and that information as to the effects of various factors, such as the pouring temperature and size of section, should accompany contraction determinations.

Of course, any casting should be designed with especial regard to foundry practice, i.e., as to the alloy to be used, the gating necessary, size of sections, pouring temperature required, and other factors. The proposed design, which should be the result of the combined knowledge of the foundrymen, pattern-maker, metallurgist, and designing engineer, should be submitted to the pattern-maker with a statement of the definite conditions of the foundry practice to be followed. If, now, in addition, accurate figures were available as to the contraction of the alloys, together with information as to the effects of various factors, such as the size of sections, pouring temperature, and method of gating, upon the contraction, then it would be a much more simple technical problem to construct a pattern that would yield castings of the required dimensions, and with a minimum number of wasters in production, than it is where a general rough value is employed for the pattern allowance for all aluminium alloys. The question of pattern allowances for aluminium alloys and contraction on freezing has been discussed by the author [52] in another place.

Castings should be designed so that the variations in thickness of contiguous sections is as small as possible, and sharp changes in direction are to be avoided, as should thin sections and complicated coring. However, where variation in the thickness of neighboring sections and abrupt changes in direction can not be avoided, ample fillets should be used. The size of the fillet will necessarily vary with the abruptness of the change in direction (angle made) and the difference in thickness of sections; larger fillet radii are required with sharper angles and greater differences in thickness. On the other hand, the fillet radius should not be too large since, otherwise, draws and shrinkage holes may occur, which the use of the fillet is intended to overcome. Practical experience will best dictate the practice to be followed here. Steel inserts are bad since they often cause blows and draws, but it is often necessary to mold such an insert into the sand and cast the alloy around it, for example, in the instance of oil pipes in motor crankcases.

Molding Sands.—The question of molding sands and core sands for aluminium-alloy founding has been studied in a preliminary way by the U. S. Bureau of Mines in cooperation with the Joint Molding Sand Research Committee of the American Foundrymen's Association and the National Research Council,[49, 50] but there remains much investigative work to be done on this subject. In a general way, it may be said that the properties desired in a molding sand, whether for iron, steel, brass, or aluminium-alloy founding, have been fairly well determined upon by foundrymen in an empirical way, based on experience, and the properties may be classed as follows:

1. Texture of the sand, or its fineness of grain. This varies, depending upon the type and kind of castings which are to be made.
2. Cohesiveness, or bonding power. A sand should bond well naturally, without the addition of bonding agents.
3. Refractoriness, or resistance of the sand to the temperature of the liquid alloy. This is not important in aluminium-alloy work.
4. Permeability, or porosity of the sand to the escape of gases given off by the alloy during pouring and freezing. This is of great importance in the case of sands for aluminium-alloy work, and sands of high permeability, i.e., readily self-venting, are desired.
5. Durability of the sand, so that it may be used over and over again without becoming " dead."
6. Ease of " shaking out " of the molds. This is desirable in sands for all alloys.

The above properties, or qualities rather, of molding sands vary depending upon the physical characteristics and the chemical composition of the sands. Of course sands may be exceedingly complex from the mineralogical point of view,[15] but, briefly, the minerals in sands may be divided into three main classes, as follows:

1. Quartz, or silica. This is the highly refractory substance of sands, and it has no bonding effect.
2. Clay, or clayey substance. This includes kaolin and related minerals, and is the bonding substance in sands.
3. Feldspar and related minerals. Feldspar is relatively non-refractory and fluxes at moderate temperatures. It is an undesirable constituent, but is present in all natural sands.

The value of any sand for molding is dependent upon the relative amounts and physical characteristics of the above materials in the sand.

The ordinary molding sands that are used for brass are satisfactory for aluminium alloys, and in the United States the well-known Albany sands are employed largely for light alloy founding. These sands are employed also for gray iron. Other sands used for aluminium alloys include those of New Jersey, known as Millvale, North River, Howell, Lumberton, and Richland. Thames sand has been imported from England for aluminium-alloy work, although needlessly since it is quite similar in texture, analysis, and behavior in the foundry to Albany sand. Similarly, sands have been imported from France, with the mistaken idea that foreign sands possess some undescribed properties which make them especially suitable for aluminium-alloy work. It is not necessary here to take up the question of domestic sources of molding sands suitable for light alloy founding, and it is sufficient to state that ample supplies of such sands are available. Table 89 gives typical analyses of two samples of Albany sand, while Table 90 gives the mineral composition of sample 2 (Table 89) calculated from the chemical analysis.

TABLE 89.—*Chemical analysis of two samples of Albany sand.*

Constituent.	Percentage composition.	
	Sample 1.	Sample 2.
Silica	81.45	75.91
Alumina	7.30	9.44
Ferric oxide	4.10	3.26
Ferrous oxide	....	1.86
Magnesium oxide	0.68	0.64
Calcium oxide	0.90	1.12
Potassium oxide	1.40	1.42
Sodium oxide	1.38	2.96
Titanium oxide	....	0.64
Loss on ignition	2.50	2.91

The Albany sands are graded in the trade according to fineness, and the fineness is expressed numerically. Thus the fineness grades of G. F. Pettinos, Philadelphia, Pa., who handles Albany sands, are 5, 4, 3, 2, 1, 0, 00, and 000, in increasing fineness. Various other methods of grading are employed by dealers and sand producers. Boswell [46] has adopted the limits given in Table 91 as to the limits of grading on mechanical analysis, and gives the data in Table 92 for eight samples of Albany sands of

TABLE 90.—*Mineral composition of Albany sand (sample 2 of Table 89) calculated from the chemical composition.*

Constituent.	Per cent.
Quartz............................	51.63
Lime..............................	5.57
Soda..............................	12.05
Potash............................	17.53
Clay substance....................	4.74
Hydrated iron oxide...............	6.13
Other constituents................	2.51
Total.............................	100.16

TABLE 91.—*Limits for grades of molding sand as determined by mechanical analysis.*[a]

Approximate mesh of screen, mm.	Symbol.	Size of grains, diameter.		Remarks.	
		Greater than, mm.	Less than, mm.		
....	G C	2.0	....	Gravel grade (G)	
1 to 25	V C S	1.0	2.0	Very coarse sand	Sand grade (S)
20 to 12	C S	0.5	1.0	Coarse sand	
50 to 20	M S	0.25	0.5	Medium sand	
120 to 50	F S	0.1	0.25	Fine sand	
180 to 120	c s	0.05	0.1	Superfine sand or coarse silt	Silt grade (s)
....	f s	0.01	0.05	Silt	
....	c	....	0.01	Clay or mud grade (c)	

[a] Based on Boswell.

TABLE 92.—*Mechanical analysis of eight grades of Albany molding sands.*[a]

Grade called.	G per cent.	V C S per cent.	C S per cent.	M S per cent.	F S per cent.	s per cent.	c per cent.	S+G per cent.
5	2.0	8.5	16.5	30.5	7.0	25.2	10.3	64.3
4	1.3	2.5	15.8	24.0	10.7	30.0	15.7	54.3
3	0.5	0.9	6.1	39.2	17.3	26.1	9.9	64.0
2	......	......	2.4	19.8	36.4	31.1	10.3	58.6
1	......	......	1.1	6.3	19.0	64.3	9.3	26.4
0	......	......	0.9	5.8	7.9	72.5	12.9	14.6
00	......	......	0.5	1.8	5.7	83.1	8.9	8.0
000	......	0.4	0.5	3.5	3.7	78.9	13.0	8.1

[a] Based on Boswell.

different grades, handled by G. F. Pettinos, Philadelphia, Pa. Table 93 gives some data on Grade 0, Albany sand, handled by Whitehead Brothers Co.*

TABLE 93.—*Characteristics of Albany molding sand, grade No. 0, of Whitehead Brothers Co.*

Chemical analysis.	Per cent.	Mechanical analysis.	Per cent.
Loss on ignition	1.95	Sand retained on 20-mesh sieve	None
Silica (SiO_2)	80.52	Sand retained on 40-mesh sieve	0.1
Iron oxide (Fe_2O_3)	4.83	Sand retained on 60-mesh sieve	0.5
Aluminium oxide (Al_2O_3)	8.39	Sand retained on 80-mesh sieve	0.6
Calcium oxide (CaO)	0.67	Sand retained on 100-mesh sieve	7.6
Magnesium oxide (MgO)	0.64	Sand retained on 150-mesh sieve	20.1
Sodium oxide (Na_2O)	1.25	Sand retained on 200-mesh sieve	16.5
Potassium oxide (K_2O)	1.45		
		Fine silt 200+	39.2
		Clay substance	15.4
		Sharp sand	84.6
		Bond	15.4
		Fineness, No. 200	

Rational analysis.	Per cent.
Clay substance	19.85
Quartz	50.50
Feldspar	29.65

Bond absorption, Total, 340, Quality, fair	Approximate composition of bond.	Per cent.
Transverse strength (oz. per sq. in.), 320	Clay (aluminium silicate)	45
	Hydrated iron oxide	55
	Organic bond	

While most of the aluminium-alloy castings made in the United States are molded in Albany sand, local sands from near Zanesville, Ohio; North Canton, Ohio; Conneaut, Ohio; in Illinois, Kentucky, Michigan, and elsewhere, are employed by a number of foundries. Newland [17, 18] and others have described in detail the famous Albany sands, and detailed information may be had from his and other reports.

In molding aluminium-alloy sand castings, the sand should be worked rather dry and the molds rammed rather lightly. The actual water content of tempered sand varies from 5 to 10 per cent, and 7 per cent of added water is suitable for average sand for most work. In large foundries laid out for straight-line production the use of mechanical sand-handling systems has been more frequent in recent years; this is due to the increased size of aluminium-alloy foundries, and the employment of modern

* Private communication, H. B. Hanley, Aug. 26, 1922.

production methods. At the same time, sand mixing and compounding of sands, for core work in particular, has been raised to a higher standard, and a number of special sand mixtures have been devised for special purposes. Wet and improperly tempered sand can be eliminated by machine mixing and careful supervision, and the preparation of sand in the right condition is an important step in the production of sand castings. So far as the available literature is concerned, there appears to have been but little scientific investigation of the difficulties met with in non-ferrous molding sand practice.[33] The non-ferrous alloys, particularly the light aluminium casting alloys, are run at rather low temperatures when compared with steel-casting practice. Burning-on difficulties, which have never been eliminated totally in steel practice, are not met with in aluminium-alloy founding, but the increased tendency to oxidation, greater mobility and sand-searching power of aluminium alloys, present peculiar difficulties. Sand penetration by mobile metal may be prevented by using sands of fine grain and close bond, though the self-venting qualities and permeability of a mold are thus sacrificed in part.

Cores and Core Sands.—So many of the wasters occurring in aluminium-alloy castings production are due to core-room practice, and so little has been done to determine the various factors which influence the preparation of a suitable core for such work, that the whole question of cores and core sands is one that requires thorough investigation. A core is a shaped body of sand placed in a mold so as to give a corresponding cavity in the casting. As a rule, cores should be stronger than molds since they are subject to more severe handling; further, since the cores are surrounded by the liquid alloy, they not only must have the strength to withstand the bending strains caused by the exerted pressure, but they must be sufficiently porous or so vented that the gas formed is given ample opportunity for escape without blowing. Moreover, since the cores must be removed from the castings on shaking out the molds, the bond should be such that it does not set hard on freezing of the alloy.

While the use of cores should be avoided wherever possible in the design of aluminium-alloy castings, much coring is necessary in complicated castings. Cores should be so made that they will crush readily and offer no resistance to the normal shrinkage of the casting. A hard core will cause strains to be set up that may cause the metal to draw apart. Hence, the cores should be light and soft, yet sufficiently hard to stand handling and trucking in the foundry. When making cores, it should be remembered that most of the light aluminium alloys have large linear contraction and are hot-short; a hard core will consequently cause cracked castings.

A large variety of sands and mixtures of sands, as well as binders, is employed, and it is often that several different core mixtures may be used in one foundry for different classes of work. Lake sands and local bank sands are used considerably for cores, and the mixtures of sands employed are so many and varied that only a few examples can be given here. Some special core sands, such as Providence river sand, are handled by foundry dealers especially for aluminium-alloy castings. The question of core sands and core-sand mixtures has been discussed by a number of workers, and a paper by Pouplin [45] is of interest. Good core sands are characterized by their cleanliness, freedom from silt and clay, their uniformity of grain, and high silica content. The texture of core sands is very variable, running from very fine to very coarse grains, and in making up core-sand mixtures the texture may be very variable because of the different sands used in the mix.

For aluminium-alloy founding, a core-sand mixture consisting of 42 lb. of sharp sand (Michigan lake sand) and $1\frac{1}{4}$ lb. of oil is used by one plant. In another, a mix of 3 parts new core sand (sharp white sand), 1 part new molding sand, 4 parts old burnt core sand is employed; this is mixed with 1 part binder (corn-flour, gluten, or molasses) to 20 or 25 parts sand. In a foundry making vacuum-sweeper castings the following mixture is used; $1\frac{1}{2}$ parts old core sand, 1 part new core sand (lake), $\frac{1}{2}$ part molding sand, and this mix is made up with 1 part binder (dextrin) to 45 parts sand. Another firm employs the following mix: 100 lb. Providence river No. 1 core sand, 1 lb. core binder (commercial), and 1 qt. boiled linseed oil. For small cores, an oil-sand mixture is favored, while for large cores sand mixtures containing molasses, resin, and flour are employed. Resin is a good binder because it will immediately soften under the heat when the casting is poured. and the core will yield readily. Moreover, such resin-containing cores can be readily knocked out of the castings, provided this is done while the castings are still hot; on the other hand, if they are allowed to cool, the resin will harden and it is then very troublesome to remove the cores.

Dextrin may be used as a binder for cores, a typical mixture being made up as follows: 1 part of molding sand and 3 parts core sand are mixed thoroughly, and then 25 parts of this mixture are used with 1 part of dextrin. A stronger core mixture may be had by using 1 part dextrin to 15 parts sand. Another typical core-sand mixture consists of 28 parts silica or lake sand, 12 parts molding sand, and 1 part linseed oil; the lake and molding sands are mixed thoroughly, adding water until the mix is almost damp enough for use, and then the linseed oil is added. Cores may be made, also, of core sand mixed with molasses in the proportion of about 15 to 1, but cores made with either linseed oil, resin, or molasses should not be allowed to cool since they will set hard, stick to the metal, and be difficult to knock out of the casting. There should not be any carbon dust or blacking used with the core, since these materials fill up the pores and prevent the gases from escaping; they also make the castings dirty.

The tendency in aluminium-alloy foundry practice is toward the use of green-sand cores wherever possible because of the greatly decreased production costs which may be effected by their employment, rather than dry-sand cores. Practice still is rather variable, some founders using dry-sand cores exclusively, while others use dry-sand cores only where necessary. Green-sand cores are used largely in the founding of large automotive castings as body cores for crankcases, oil pans, and transmission housings, and green-sand cores are used to an increasing extent in casting cooking utensils. The accuracy of castings with particular regard to hubs, bosses, and other attachments in crankcases for example, is more likely with green-sand cores rather than dry-sand ones, since the slightest swelling of the dry-sand core, which is difficult to prevent, is accentuated in the casting, and the bosses can not be centrally drilled. Molded in green sand, castings are liable to be more free from concealed cracks, produced by the unyielding nature of the dry-sand cores. The economy of the green-sand core method is also apparent since a large number of core-makers are dispensed with, and the cost of drying cores is eliminated. In vacuum-cleaner work, dry-sand cores are generally preferred, although green-sand cores are being employed to an increasing extent. In some foundries, green-sand cores are used for heavy, chunky castings and dry-sand cores for light, open jobs. Dry-sand cores are used practically exclusively for some kinds of automotive castings such as manifolds, pistons, and carburetors, particularly on small complicated jobs which are difficult to run, and green-sand cores are employed where practical in large castings.

Molding Methods.—The fundamental principles governing the molding of aluminium-alloy castings are broadly the same as those for brass and bronze, but necessarily different in certain marked essentials. Normally, the gases evolved in casting an alloy into a sand mold must be drawn either through the mold or through the metal; if they go through the metal rather than the mold, a blowholey casting will result. Owing to the low specific gravity of the aluminium alloys, the gases will readily tend to escape through the liquid metal rather than through the sand of the mold, unless the latter is sufficiently porous. Hence, porosity is one of the cardinal requirements of a mold.

Numerous rules have been given for the molding and casting of aluminium alloys, some of which are highly contradictory. In a very general way, without taking up the details of aluminium-alloy molding practice, it may be said that in casting aluminium alloys in sand molds, one is dealing with alloys of low specific gravity; they are also hot-short. Mold pressures are consequently not so serious as in iron and steel or brass, due to the head of metal on pouring.[21] The sand should be worked rather dry, and the molds rammed lightly. The shrinkage of the aluminium alloys is rather high, as mentioned previously, and the patterns must be made according to the determined shrinkage for a given alloy, rather than on the basis of the usual figure, 0.156 in. per ft., used so generally. For large castings, such as automotive crankcases, aluminium-alloy molding boxes are now employed quite generally, and these boxes are convenient to handle and reasonably inexpensive. For ordinary small sand castings, snap flasks are desirable, since after the mold is completed, the flask can be removed, and the mold placed on a flat board and then covered with a suitable board so that the pouring gates are exposed. This can be done because of the low specific gravity of the alloys, and it effects a saving of flasks.

Ramming.—The porosity required in molds can be obtained, of course, by using very dry sand; if this be employed, however, foundry difficulties are introduced, such as drops and crushes, or if these do not occur, cracks may be caused on account of the unyielding quality of the dry sand. A properly tempered green sand should be used, and it should be *lightly* but evenly rammed. The amount of ramming given to a mold is one of the principal items of difference in the founding of aluminium alloys and brass or steel. Since the specific gravity of aluminium alloys is low, the head pressure is practically negligible; hence, it is not necessary to ram the mold so hard as for brass in order to maintain the shape of the pattern. Light ramming will provide the necessary porosity, hence the sand may be tempered about the same as for the other non-ferrous metals.

Even ramming is also a *desideratum*, and the easiest and fastest way of accomplishing both even and light ramming is by means of power-molding machines. Jar- or jolt-rammed rollover molding machines are much used in the molding of crankcases, oil pans, and other large castings. In hand-rammed molds, the skill of the individual molder must be largely depended upon. In machine ramming, any required compactness of the mold may be secured by varying the number of jolts, and it will be also certain that the mold is rammed evenly throughout. After a mold has been rammed and the pattern drawn, it may be dusted over with lycopodium powder or French chalk, and the cores then set where cores are necessary. Plumbago and other black facings should not be used in molds where a white clean surface on the castings is required. The molding of a crankcase will be described in later paragraphs of this chapter showing the use of machine molding on large work.

Gating and Risers.—The importance of overcoming shrinkage in casting has already been mentioned, and the difficulties incident to shrinkage may be surmounted largely by proper gating. Large and complicated castings are always best run from a bottom gate, since this gives a more even flow of metal and prevents the displacement of cores. The use of large gates and risers [37] is generally recommended, so that the metal will not be restricted in its flow into the mold. Risers must be ample in size since otherwise they will cool before the casting. The principles of gating in the molding of aluminium-alloy sand castings are the same as for other alloys, but the hot-shortness, high shrinkage, and low specific gravity of the aluminium alloys make it necessary that these principles be carefully considered. Shrinkage may be considerably eliminated by so arranging the gates and the distance traversed by the metal through the mold (before arriving at its final destination) so that all parts solidify at the same time. As is known, the strength of a casting varies from section to section, and shrinkage is undoubtedly the main cause of weak sections. This will be readily apparent, because if one section of a casting freezes much before a neighboring section, the first section will be fed by the second one, and insufficient metal will be left available to supply the second section. This can be eliminated largely, or in part, by two methods of molding: First, the gating may be so arranged that when the mold cavity is full of metal, the length of the path traversed by the metal will have been such that the heavy sections will be filled with colder metal than the light sections. On account of mechanical difficulties in molding, this method does not have any broad application. Second, risers may be used.

Since the light aluminium alloys must be poured rapidly, it is advisable to have larger gates than are normally used for brass or bronze; the size of the gate is determined by the size and shape of the casting, as well as the method of gating. The position of the gate also depends upon certain factors. Thus, generally, castings should be gated at the end, rather than at the middle, and gating at the bottom is good where possible. If a gate is attached to a heavy section that is contiguous to a light one, draws and internal holes may appear in the heavy section; in this case, it would be preferable to gate to the light section and control the rate of solidification of the heavy section by means of a chill. This can often be done advantageously unless the difference in the masses of the light and heavy sections is too great. Where there are large differences in the mass of adjoining sections, a riser should be attached to the heavy section. Heavy risers should be used on aluminium-alloy castings, and these should be attached to those parts of the casting that are the last to solidify so as to feed the parts. Risers should be of such a size and so situated that they will remain liquid for a much longer time than the casting. Risers are not necessary always on small castings, but on large castings and on heavy parts of castings they are most necessary. Molds should be so gated and the risers attached in such a way that the minimum pouring temperature can be employed.

The use of shrinkage balls in aluminium-alloy casting practice has been generally applied to all castings wherever possible in one foundry, with good results in reducing casting losses arising from shrinkage holes and cracks.

Use of Chills.—Chills are used frequently in the molding of aluminium-alloy sand castings, particularly in places where there are large lugs, bosses, or other masses of metal contiguous to light thin sections. Chills may be made of iron, brass, or aluminium. In general, it is better practice to use risers instead of chills where possible, i.e., if the riser will accomplish the same result. The tendency is to avoid the employment of specially shaped chills and use risers, provided that the molding of the riser and its removal in the chipping room do not cause too much difficulty. Of course, the individual casting should be carefully examined before deciding between risers and chills, but as a rule it is preferable to employ a riser that can be molded readily, and subsequently easily cut off, than to use a special chill or a standard chill at a position in the mold where it will be difficult to support, or where it will spoil the appearance of the casting. However, it is at times more desirable to use a standard chill where it can be readily supported in the mold, rather than a riser which is difficult to mold and that will be troublesome to remove by trimming. It may be pointed out that while both chills and risers have definite advantages and disadvantages, there is a definite element of cost involved in their use. Often it is possible to use chills instead of risers and thereby gain both in time and labor.[4]

CRANKCASE FOUNDING.

While the details of molding methods for various types of castings are of interest, it is not necessary to describe these at length here, but it may be found of interest if machine molding in the production of a motor crankcase is discussed briefly. Figs. 121 to 126, inclusive, show the various steps in the molding of the crankcase for the 12-cylinder Liberty aviation engine. Fig. 121

518 *Metallurgy of Aluminium and Aluminium Alloys*

shows an Osborn jar-rammed roll-over molding machine with the pattern mounted and a completed mold at the right, which has been removed from the machine. *A* indicates the locking lever

FIG. 121.—*Pattern mounted on roll-over machine in crankcase founding (The Osborn Manufacturing Co.).*

used during the roll-over operation and *B* and *B* are horn gates used for this particular casting. Fig. 122 shows a step in the process of making the mold. The flask has been placed around the pattern, the facing sand has been riddled in, and a small quantity of molding sand dropped from one of the overhead bins. Three of the molding operators are working on the pattern, and

FIG. 122.—*Molding a crankcase on a roll-over machine (The Osborn Manufacturing Co.).*

the fourth is ready to drop more sand into the flask from the overhead hopper. Two overhead hoppers are used for speed in adding sand. The next operation is jolt ramming, which is accomplished by the machine, requiring 30 secs.; air power is used. The mold is then butted off, and the bottom board applied, the mold rolled over on the machine and deposited upon the leveling car, which has been pushed into place underneath the roll-over table. The pattern is drawn in the next operation by means of the vibrator, and the finished mold is then

FIG. 123.—*Finished drag mold for crankcase before setting cores* (The Osborn Manufacturing Co.).

pushed from underneath the table into the position shown in the right of Fig. 121. Fig. 123 shows a close-up view of the finished drag before the cores are set. The vertical walls and intricate surface of the mold may be noted; the pattern is drawn so accurately by the machine that no patching or slicking up of the mold is subsequently required. The horn gates have been drawn, leaving the four gates shown on the sides of the mold; these are fed by a runner in the cope, which is practically a flatback. Fig. 124 shows a view of the drag of Fig. 123 after all the cores have been set.

Fig. 125 is a view showing a floor of molds, and indicates the continuous pouring system used; in the foreground are shown the drags as received from the molding machine. Cores are to be set into the last drag in the line, shown in the right foreground. The core is shown with the lifting handles in position. Farther

FIG. 124.—*Finished drag for crankcase after setting cores (The Osborn Manufacturing Co.).*

back along the row of molds the cores have been set in the drags, while farther back the copes have been put on and the molds closed. Still farther back molds have been poured, and at the extreme end they have been shaken out. Fig. 126 shows the finished crankcase after trimming of runners and gates, and cleaning.

The production data on this job may be found of interest. With a crew consisting of four molders, eleven helpers, and eleven laborers, who do all the molding, core-making, core-setting, closing, pouring, shaking out, and flask handling, 102 castings are

522 Metallurgy of Aluminium and Aluminium Alloys

FIG. 125.—*View of line of molds for continuous pouring in crankcase founding (The Osborn Manufacturing Co.).*

FIG. 126.—*Finished crankcase, upper half (The Osborn Manufacturing Co.).*

made per day, or an average of four castings per man per day.*
In comparing this with hand molding, the fact must not be lost
sight of that while it might be possible for a man to mold four of
the castings in a day, it would not be possible for him to do all the
work in connection with the production thereof, as is accomplished
by a machine crew.

CLEANING, CHIPPING, AND GRINDING CASTINGS.

After a rough casting is shaken out of the mold, the core
sand is removed, and this may be done by loosening it with
mechanical vibrators or with hand rammers. The gates and
runners are partly knocked off by hammers and, in part, cut off
on band saws, depending upon the casting and whether it is
handled hot or cold. All fins, gate lugs, and other protuberances
may be removed by chipping or grinding (cf. also Chapter XIX).
Chipping, by means of pneumatic chipping chisels, is employed
for cleaning crankcases, oil pans, and other large castings, but
small castings are ground on grinding wheels. Fig. 127 is a
view in the cleaning room of a foundry. After chipping and
cleaning, the castings may be sand blasted or not, and they are
then ready for shipping or machining. In one large foundry all
kinds of aluminium-alloy castings from small cooking utensils to
large crankcases are blasted with flint shot. The work is done
on a conveyor table in a closed room. The castings are sent in
through an opening in one end of the room, slide along the rollers
of the conveyor table, and after being cleaned by blasting with
flint shot are sent out through an opening in the other end of the
room to the inspection and cleaning departments. According
to report, for a large crankcase, about 9 oz. of flint shot is used,
the work being done under a pressure of 105 lb. per sq. inch.
The practice in cleaning of aluminium-alloy automotive castings
at the plant of the International Motors Co., Brunswick, N. J.,
is described by Price;[55] flint shot is used for sand blasting.
After blasting, the castings are ground on hand grinders for
the purpose of snagging off small sharp protuberances, or on
machine grinders for flat surfaces. Large castings are usually
ground, but small parts may be filed in preference to grinding.

---

* Private communications, E. S. Carman, July 26, and Aug. 13, 1920.

Fig. 127.—View in cleaning room of a foundry.

SELECTED BIBLIOGRAPHY.

Numerous references are given in the trade journals to aluminium-alloy foundry practice, and many descriptions have been published in these journals of typical foundries. The literature as to the more technical aspects of founding is rather sparse, however, and in the following bibliography there are included most of the important references.

1. Anon., The use of aluminium in automobile castings, *The Foundry*, vol. 22, 1903, pp. 255–256.
2. Anon., The use of aluminium in automobiles, *The Metal Ind.*, vol. 1, 1903, pp. 5–6.
3. Ries, H., and Gallup, F. L., Report on the molding sands of Wisconsin, Geological Survey of Wisconsin, report for 1905.
4. Brown, A., The use of chills in making aluminium castings, *Brass World*, vol. 2, 1906, pp. 399–401.
5. Ries, H., and Rosen, J. A., Foundry sands, Geol. Survey of Michigan, report for 1907.
6. Lake, E. F., Aluminium castings, *The Foundry*, vol. 30, 1907, pp. 160–165.
7. Fawsitt, C. E., Viscosity determination at high temperature, *Jour. Soc. Chem. Ind.*, vol. 93, part 2, 1908, pp. 1299–1307.
8. Carpenter, H. C. H., and Edwards, C. A., The production of castings to withstand high pressures, Proc. Inst. of Mech. Engrs., vol. 74, 1910, pp. 1597–1634.
9. Gilbert, A. H., Designs of aluminium castings, *The Automobile*, vol. 24, 1911, pp. 1342–1344.
10. Gillett, H. W., and Skillman, V., The pyrometer in the aluminium foundry, Trans. Amer. Brass Founders' Assoc., vol. 5, 1911, pp. 63–69.
11. Gillett, H. W., The influence of pouring temperature on aluminium alloys, Eighth Intern. Cong. Appl. Chem., vol. 2, 1912, pp. 105–112.
12. Anon., Use of green sand cores in aluminium work, *The Foundry*, vol. 40, 1912, pp. 141–146.
13. Anon., The advance of aluminium in the foundry, *The Metal Ind.*, vol. 10, 1912, pp. 489–493.
14. Anon., Producing castings for the automobile trade, *The Foundry*, vol. 41, 1913, pp. 87–92.
15. Condit, D. D., Molding sand tests. Mineral characteristics of the molding sands, Trans. Amer. Foundrymen's Assoc., vol. 21, 1913, pp. 21–27.
16. Collins, J. W., Suggestions for making aluminium castings, *The Foundry*, vol. 42, 1914, pp. 67–68; abst. of paper before the Detroit Foundrymen's Assoc.
17. Newland, D. H., Albany molding sand, Twelfth report of the Director, New York State Museum, Bull. 187, 1915, pp. 107–115.
18. Newland, D. H., Albany molding sand, Trans. Amer. Foundrymen's Assoc., vol. 24, 1916, pp. 161–176.
19. Clarke, R. R., Gating non-ferrous metal castings, Trans. Amer. Inst. of Metals, vol. 10, 1916, pp. 25–46.
20. Reardon, W. J., Foundry problems, *The Metal Ind.*, vol. 14, 1916, pp. 155–158.
21. Lane, H. N., Functions of sand binders, Trans. Amer. Foundrymen's Assoc., vol. 24, 1916, pp. 192–200.
22. Anderson, R. J., The practice of melting and casting aluminium, *The Foundry*, vol. 46, 1918, pp. 104–106; 164–166.

23. Hill, E. C., Aluminium: its use in the motor industry in England, *The Metal Ind.*, vol. 16, 1918, pp. 543–546.
24. Boswell, P. G. H., A memoir on British resources of refractory sands for furnaces and foundry purposes, Part I, Taylor and Francis, London, 1918.
25. Anon., Progress in aluminium casting, *The Metal Ind.*, vol. 17, 1919, pp. 211–213.
26. Anon., Attaining production on Liberty motor crankcase castings, *The Foundry*, vol. 47, 1919, pp. 126–129.
27. Anon., Aluminium foundry molding losses analyzed, *The Foundry*, vol. 47, 1919, p. 416.
28. Anderson, R. J., Unsoundness in aluminium castings, *The Foundry*, vol. 47, 1919, pp. 579–584.
29. Lea, F. C., Aluminium alloys for aeroplane engines, *The Aeronautical Jour.*, vol. 23, 1919, pp. 545–607.
30. Anderson, R. J., Blowholes, porosity, and unsoundness in aluminium-alloy castings, U. S. Bureau of Mines Tech. Paper 241, December, 1919, 34 pp.
31. Anderson, R. J., Metallography of aluminium ingot, *Chem. and Met. Eng.*, vol. 21, 1919, pp. 229–234.
32. Anderson, R. J., Special and commercial light aluminium alloys, U. S. Bureau of Mines War Minerals Investigations Series No. 14, April, 1919.
33. Boswell, P. G. H., Molding sands for non-ferrous foundry work, *Jour. Inst. of Metals*, vol. 22, 1919, pp. 277–298.
34. Carman, E. S., Machine molding in brass and aluminium foundries, *Brass World*, vol. 15, 1919, pp. 357–358.
35. Anon., British aluminium foundry practice, *The Metal Ind.*, vol. 18, 1920, pp. 163–165; 221–224.
36. Prentiss, F. L., Modern foundry for aluminium castings, *The Iron Age*, vol. 103, 1920, pp. 535–539.
37. Gibson, W. A., Position of tensile tests in the foundry, *The Iron Age*, vol. 105, 1920, pp. 725–728.
38. Anderson, R. J., and Capps, J. H., Investigation of hard spots in aluminium, *The Foundry*, vol. 48, 1920, pp. 337–342.
39. Wolf, F. L., and Grubb, A. A., Laboratory testing of sands, cores, and core binders, Trans. Amer. Inst. of Min. and Met. Engrs., vol. 64, 1920, pp. 630–637.
40. Anderson, R. J., Analysis of losses in aluminium shops, *The Foundry*, vol. 48, 1920, pp. 969–992; and *idem*, vol. 49, 1921, pp. 16–21; 54–57; 104–111; 143–147; 188–191; and 235–239.
41. Anderson, R. J., Casting losses in aluminium-foundry practice, Trans. Amer. Foundrymen's Assoc., vol. 29, 1921, pp. 457–489.
42. Foote, P. D., Fairchild, C. O., and Harrison, T. R., Pyrometric practice, U. S. Bureau of Standards Tech. Paper No. 170, February 26, 1921.
43. Turner, T., The casting of metals, *Jour. Inst. of Metals*, vol. 26, 1921, pp. 5–43.
44. Grubb, A. A., and Jamison, U. S., Air required in baking cores made with linseed oil, *Chem. and Met. Eng.*, vol. 25, 1921, pp. 793–795.
45. Pouplin, G., Core sand in the foundry, *La Fondrie Moderne*, trans. by P. Powell, in *Brass World*, vol. 17, 1921, pp. 345–350.
46. Boswell, P. G. H., A comparison of British and American foundry practice with special reference to the use of refractory sands, The University Press of Liverpool, Ltd., London, September, 1922.
47. Anderson, R. J., Inclusions in aluminium-alloy sand castings, U. S. Bureau of Mines Tech. Paper 290, June, 1922, 25 pp.
48. Anderson, R. J., Cracks in aluminium-alloy castings, Trans. Am. Inst. of Min. and Met..Engrs., vol. 68, 1923, pp. 833–860.

49. Anderson, R. J., Reclamation and conservation of sands for molds and cores in aluminium-alloy foundry practice (Report to the Joint Molding Sand Research Committee), February, 1923.
50. Anderson, R. J., and Dalrymple, G. B., Compilation of questionnaires sent out to aluminium-alloy foundries on the use of sands for molds and cores in light alloy work (Report to the Joint Molding Sand Research Committee), April, 1923.
51. Guillet, L., Un nouvel alliage d'aluminium: l'alpax, *Le Génie Civil*, vol. 82, 1923, pp. 413-419; 441-444; and translation by R. J. Anderson, paper before Amer. Foundrymen's Assoc., Milwaukee meeting, Oct., 1924.
52. Anderson, R. J., Linear contraction and shrinkage of a series of light aluminium alloys, Trans. Amer. Foundrymen's Assoc., vol. 31, 1924, pp. 392-466.
53. Mills, Jr., W. A., Casting automobile aluminium radiator shells, Trans. Amer. Foundrymen's Assoc., vol. 31, 1924, pp. 467-468.
54. Hurren, F. H., The influence of casting temperature on aluminium alloys, *The Foundry Trade Jour.*, vol. 29, 1924, pp. 75-79.
55. Price, B. K., Cleaning automobile castings, *The Foundry*, vol. 52, 1924, pp. 355-357.
56. Simonds, H. R., Casting aluminium cycle parts, *The Foundry*, vol. 52, 1924, pp. 661-664.
57. Anon., Conveyor system in aluminium foundry, *The Iron Age*, vol. 114, 1924, pp. 843-844.
58. Knerr, H. C., Aluminium-alloy castings from sheet scrap, paper before Amer. Foundrymen's Assoc., Milwaukee meeting, Oct., 1924.
59. Basch, D., and Sayre, M. F., Foundry treatment and physical properties of silicon-alumininium sand castings, paper before the Amer. Foundrymen's Assoc., Milwaukee meeting, Oct., 1924.

CHAPTER XII.

CASTING LOSSES, DEFECTS IN CASTINGS AND THEIR PREVENTION.

The study of wasters in castings production and defects in castings is extremely interesting from the metallurgical point of view, and the subject is of great importance to foundrymen because of its monetary aspect. Wasters in the production of aluminium-alloy sand castings (as well as die castings and permanent-mold castings) are high, and a serious source of financial loss. Taken by and large, men usually prefer to say nothing about their failures, and that a detailed description of them should be prepared would ordinarily be regarded as a waste of time. Hence, although most foundrymen are well acquainted with the causes for defects in castings, there are not many foundries, especially in the aluminium-alloy field, where systematic methods for classifying and recording defects are in force. During the World War the wasters incurred in the production of aviation engine castings were so numerous that motor production was hampered, and it became necessary to make detailed investigation of the subject. In 1918, study was made of losses in the production of aviation-engine castings in the United States by the present author for the Bureau of Aircraft Production, and detailed investigation was made later of the losses incurred in aluminium-alloy founding in general by the author in the U. S. Bureau of Mines. The results have been reported in published papers.[27, 28, 35]

Due to inherent conditions in the aluminium-alloy foundry industry, viz., the large number of small shops, the variety of castings made, and general lack of metallurgical supervision, casting losses are normally high on the average, when referred to the best practice. In this industry at least $600,000 per year could be saved if the present average casting losses were reduced 50 per cent, and this can be readily accomplished if only the preventable defectives are avoided. Of course, defective castings are not total loss, since they possess a residual scrap value for re-melting, but if much machine work is put upon a casting before it is eventually scrapped, the cost of defectives becomes large. It is thoroughly recognized by progressive founders that anything which can be done to reduce casting losses at reasonable expense is well worth while. Appreciable, and, in many instances, readily avoidable losses have occurred in the production of aluminium-alloy castings for many years, and the average of loss in founding these castings is generally much higher than in brass, bronze, cast-iron, and steel work. Thus, in the production of aviation-engine castings, the rejections on the foundry floor were as high as 30 per cent, and more at times, and for considerable periods, while the average casting loss in founding automotive and other castings is 10 to 15 per cent.

Aluminium-alloy foundrymen are well aware of the fact that unaccounted for difficulties in production arise from time to time. Thus, a foundry may run for a considerable period, say three months, turning out satisfactory castings with a reasonably low percentage of defectives, say 5 per cent. Suddenly, and for no particular reason that can be determined, the castings commence to run " bad," and losses mount to high figures—25 per cent or more. Casting losses may be high for several days to two weeks, when the losses will decrease and production again become normal, still for no particular reason that can be ascertained. In this connection, some founders have been in the habit of ascribing their troubles to the " metal " instead of attempting to determine whether foundry practice might be at fault. There have been difficulties directly traceable to the quality of the aluminium in the past, and there is no doubt that there will be more difficulties due to the metal in the future, but many casting losses can be traced to readily avoidable causes in the foundry practice. There are a large number of factors in practice over which the foundryman has direct control, and it is only by studying the defects occurring in castings that these can be traced to their causes.

Regarding actual wasters and defectives arising in the production of most castings, it appears that at least 75 per cent can be traced to the molding department and the remainder to metallurgy, melting, and miscellany. Rejections vary considerably in different foundries, depending upon the type of casting and local conditions. Furthermore, different castings are rejected for different reasons; porosity and leaks can not be tolerated in crankcases, manifolds, and carburetors, but these same defects are not serious in clutch cones, some kinds of housings and carriers, and rocker arms. On the other hand, surface sand holes are sufficient cause for rejection in many kinds of castings that must be highly polished, such as vacuum-cleaner housings and parts, wheel-hub caps, and instrument frames, but they may not be serious in most castings not requiring polishing. The difficulties in tracing defects to their causes are by no means small if all factors are considered, since several different causes may result in the same defect. On the other hand, it is possible to formulate a reasonably systematic classification of causes for defective castings, and the proper records can be kept by the inspection department. While it is held that accurate scrap-loss records should be kept by all foundries, and although it is believed that a careful inspection of defective castings and subsequent diagnosis of the losses will assist materially in reducing them, it can not be held that it is possible to eliminate defective castings completely in aluminium-alloy founding. It will be possible to reduce losses as much as 50 per cent and more in many foundries, but inherently some losses necessarily occur.

In connection with investigations of casting losses in aluminium-alloy founding carried out by the author, several studies of the causes for and prevention of specific defects have been completed, notably on blowholes, porosity, and unsoundness,[20, 25] inclusions,[26, 38] and cracks.[37] These types of defects are the most serious and troublesome encountered in practice, and most of the losses which can not be traced to careless or molding technique can be accounted for by these defects. The subject of casting losses and defects in castings can not be treated in detail here, and in the present chapter it is possible to give only general consideration to the subject and to discuss some particular defects. Accordingly, the following items are taken up: (1) casting losses; (2) blowholes, porosity, and unsoundness; (3) inclusions and "hard spots;" (4) cracks and related defects; and (5) salvage of defective castings.

CASTING LOSSES IN SAND PRACTICE.

Defective castings can be classified readily according to the kind and degree of the defects, but it is more difficult to determine exactly why a certain defect occurred. The castings produced during a day's run in a foundry may be divided into (1) good castings; (2) castings with remediable defects that can be salvaged by welding, soldering, or other repairing; and (3) totally defective castings. It is, of course, difficult at times to draw sharp lines of demarcation among good castings, those with remediable defects, and scrap castings, and this makes inspection a matter calling for experience and good judgment. Whether any casting is accepted or rejected depends upon the kind and degree of the defects, upon the use to which the casting is to be put, the extent and character of the subsequent machining in relation to the defects, as well as upon the rigidity of the specifications. A defect that may be so serious as to cause one type of casting to be thrown out may be regarded as trivial or of no consequence in another casting.

Typical Defects Causing Losses.—Castings may be rejected for defects resulting from one or a combination of causes including cracks; sand holes; chill blows and core blows; cold shuts; hard inclusions (hard spots); porosity and general unsoundness; blowholes; run-outs and mis-runs; uneven walls and cut-throughs, often due to core shifts; breaks in trucking, handling, chipping, cleaning, and welding; draws; and such defects as are traceable to hard ramming and soft ramming, dirty and broken cores, sand or paste in the cores, crushed molds, washed-in gates, poor patching, wet sand, broken molds, cope drop, and other items. The regulation of shrinkage and the prevention of cracks are two of the more serious difficulties in the production of aluminium-alloy sand castings, and rejections because of shrinkage and cracks are heavy on the average. Blowholes, porosity and unsoundness, hard inclusions, and cracks are the most common and important defects, and these items are discussed in later sub-divisions of this chapter.

Draws are common defects, and are closely associated with cracks, being due to some of the factors that cause cracks. Core blows and chill blows are fairly common defects; core blows are often due to hot molding sand and cold cores, while chill blows are due to cold, wet, and damp chills and hot sand. Losses resulting from broken castings, whether broken in handling, trucking, chipping, cleaning, or welding, are normally high when the production is rushed. Some castings are broken by rough handling, and this is avoidable. If castings are shaken out of the molds when too hot, breakage will be high. The percentage of castings rejected because of occluded sand particles, surface sand holes, and rough surfaces due to sand, is fairly large. Defects that can be traced to the sand are largely due to molding technique, rather than to the quality of the sand. Loose sand in the mold will give rise to surface sand holes and rough-

FIG. 128.—*Surface sand hole in vacuum-cleaner nozzle; about* $\times$ 2.

ness, and such sand may be due to wash on pouring, inadequate cleaning, or to dirty cores. Fig. 128 shows a typical surface sand hole in a vacuum-cleaner nozzle in 92 : 8 aluminium-copper alloy. Cold shuts are due simply to too low pouring temperature, and are avoidable. Short pours are due to mistakes in calculating the amount of alloy necessary to fill a mold. Run-outs are caused by improper clamping in the case of large boxes, sand under the flask joints, and insufficient weights on the mold where weights are necessary.

Core trouble gives rise to many scrap castings. Core shifts are frequent in the production of many types of castings, and are due to improper placing, sand under the prints, cores getting loose from their anchors, lack of chaplets, etc. Core shifts result in uneven walls, thin walls, and cut-through castings; they are largely avoidable. Cores that are too hard or too soft may be traced to faulty core making, and defects due to poor cores can not properly be charged to the floor molder. In general, cracks are more likely to be prevalent when dry-sand cores are used. Some defects are due to crushed and broken molds, such breakage occurring on setting cores, by improper clamping, when the bottom board [4] is not placed correctly (not rubbed down), by the mold sagging because of dry sand, and by core sags. Very small crushes will result in scrap castings, and defects due to this cause are largely avoidable. Unexpected collapses of the cope, and partial cope drop when the cope is set on, invariably ruin the mold, and drops can not usually be detected until after the casting has been poured. Blowholes, porosity, and unsoundness are frequent defects, as has been previously observed, and these defects are markedly affected by the melting temperature (because of increasing gas absorption with increasing temperature) and pouring temperature, as well as by the length of time of melting. These defects are discussed at some length in a later sub-division of the chapter. The presence of hard inclusions in aluminium-alloy castings is not ordinarily observed during the routine inspection of castings in the foundry, and complaints because of this defect usually originate in the machine-shop. Hard inclusions, i.e., the so-called hard spots of foundry parlance, are due largely to dirty melting practice, and this type of defect is discussed at greater length later.

Table 94 is a summary giving the causes for losses in the production of a variety of automotive castings in a large foundry, based on analysis of casting-loss records and an output of 12,000,000 lb. of finished castings over a 12-month period. In examining the figures, it is interesting to note that cracks accounted for the largest percentage of scrapped castings. Cope drops causing sand holes, and sand left in the molds through improper cleaning, accounted for 1.83 per cent of the rejections. Mis-runs and run-outs resulted in a loss of 2.18 per cent. Breakage in handling was 1.76 per cent, a high figure. In the classification of losses according to groups, as given by the analysis, 2.3 per cent was scrapped because of core troubles, such as cut-throughs, crushes, broken cores, core blows, cores set wrong, and core shifts, of which cut-throughs, broken cores, and cores set wrong (or shifts) caused the heaviest losses. Rejections due to molding errors contributed 3.22 per cent of the total losses, of which the greatest number of rejections were due to cope drops, sand holes, and hard or soft ramming. The other causes were responsible for the remaining losses, 8.4 per cent.

TABLE 94.—*Classification of causes for defective castings in a large foundry.*

Causes for losses.		Per cent of castings scrapped.	Totals.
Due to cores	Cut through	0.75	
	Crushed	0.21	
	Broken	0.57	
	Set wrong or shift	0.41	
	Blows	0.12	
	Miscellaneous	0.24	
	Total	2.30	2.30
Due to molding	Blows	0.24	
	Drops or sand holes	1.83	
	Chill blow	0.10	
	Hard and soft ramming	0.47	
	Wet sand	0.15	
	Chill drop	0.15	
	Miscellaneous	0.28	
	Total	3.22	3.22
Due to other causes	Cracked	2.27	
	Cracked in welding	0.36	
	Broken in handling	1.76	
	Broken in welding	0.10	
	Broken in chipping	0.34	
	Broken in knocking out cores	0.17	
	Broken in trucking	0.02	
	Warped or not to gage	0.46	
	Poured short	0.17	
	Mis-run or run-out	2.18	
	Faulty metal	0.34	
	Miscellaneous causes, not classified	0.23	
	Total	8.40	8.40
	Grand total		13.92

The whole problem of defects in aluminium-alloy castings and the factors affecting their occurrence has been discussed at great length by the author in published papers,[20, 25, 26, 27, 35, 37, 38] which may be consulted for detailed information.

Casting Losses and Defects in Castings

Factors Affecting Casting Losses.—The production of any casting starts in the drafting room, and a great many defective castings are made on the drawing board. Design is a factor over which the founder has little, if any, control, but many losses can be prevented if the foundry foreman, molder, and coremaker are called in when the question of design arises. Perkins [9] has outlined five essentials to ensure low casting losses in founding, and these will bear repetition here, viz., (1) a design from which good castings can be made continuously; (2) the combined ideas (coremaker, molder, cleaning-room foreman) of all interested in the making of the casting, so as to obtain the best equipment for its production; (3) a substantial incentive offered for good work; (4) having made a good casting, to ensure its shipment in as good condition as when it left the foundry; and (5) a clean, well-equipped shop. In making a detailed analysis of the casting losses in any foundry it is necessary to consider carefully the factors affecting these losses. The factors that affect the percentage of castings rejected for whatever reasons may be divided, for convenience, into four main classes: A, metallurgical factors; B, molding factors; C, miscellaneous factors, and D, the human element. Under these various classes, various factors fall automatically. A classification of the variable factors which, alone or combined, affect casting losses in aluminium-alloy founding is given below, but the confines of space prevent discussing them at all here.

While technical control is important, the problem of obtaining good, sound castings is mainly dependent upon skillful molding technique, adequate supervision, and the correct equipment, and the human element must be regarded as a most important factor in castings production. The subjoined figures for casting losses in a large modern foundry turning out a six-cylinder motor crankcase may be of interest in this connection:*

Proportion rejected and causes.	*Per cent.*
Castings rejected, maximum.............	29
Castings rejected, minimum.............	3
Castings rejected, average..............	9
Castings rejected for defects that could be eliminated only with great difficulty....	5
Castings rejected for defects due to carelessness, lack of supervision, etc.........	4

---

* Private communication, June, 1919.

536   *Metallurgy of Aluminium and Aluminium Alloys*

The defects that could be eliminated with great difficulty, if at all, were said to be attributable to foundry peculiarities and the human element.

Given proper design, the following factors may be regarded as intimately connected with losses in production:

A. Metallurgical Factors:
1. Kind and quality of the melting charge:
    (a) Chemical composition of the alloy.
    (b) Quality of the aluminium pig.
    (c) Method of introducing the additive elements, for example, copper.
2. Melting practice:
    (a) Kind of furnace and furnace atmosphere.
    (b) Time and temperature of melting.
    (c) Melting ratio.
    (d) Composition of the containing vessel.
    (e) Kind and quality of the fuel used.
    (f) Use of fluxes.
3. Pyrometry:
    (a) Melting temperatures.
    (b) Pouring temperatures.
4. Metallurgical specifications.

B. Molding Factors:
1. Molds and methods of molding:
    (a) Kind and character of the molding sand.
    (b) Power machine, hand machine, floor, and bench molding.
    (c) Patterns.
    (d) Flasks and boxes.
    (e) Kind of molds made for different castings.
    (f) Methods of gating, position and number of risers, and venting.
    (g) Handling and placing molds.
2. Cores and core setting:
    (a) Green-sand and dry-sand cores.
    (b) Core making.
    (c) Handling cores in the foundry.
    (d) Core setting and coring-up.
3. Inspection of molds before closing.

C. Miscellaneous Factors:
1. Total output of castings.
2. Kinds of castings made.
3. Weights of castings.
4. Pouring practice.
5. Shaking out.
6. Handling and trucking castings.
7. Grinding, chipping, and cleaning.
8. Welding and repairing.
9. Inspection of castings.
10. Machine-shop returns.

D. Human Element and Labor:
1. Quality of the labor.
2. Piece-work and day-work rates.

Losses in the Production of Castings.—Losses owing to rejections of castings because of defects therein vary markedly in different foundries, depending upon numerous factors and local conditions. According to data reported to the U. S. Bureau of Mines and from other sources, the extremes of 1 and 75 per cent have been given as the low and high losses in the production of a variety of castings. Table 95 gives a summary of the casting losses incurred in the production of a number of typical castings in three foundries. In this table, the indicated losses,

TABLE 95.—*Casting losses incurred in the production of typical aluminium-alloy castings at three foundries.*

Kind of casting.	Made in foundry.	Indicated loss, per cent.	Probable loss, per cent.	Principal causes for rejections.
Crankcase	A	11.73	12	Cracks, crushed molds, broken in handling and chipping.
Crankcase	B	4.55	7	Sand holes.
Crankcase	C	7.68	8	Core shifts and cracked in welding.
Oil pan	B	7.43	10	Sand holes and chill blows.
Oil pan	B	4.53	7	Sand holes and mis-runs.
Oil pan	B	3.47	6	Sand holes.
Oil pan	C	11.74	12	Core shifts and cracks.
Manifold	B	38.35	40	Core shifts and core blows.
Manifold	C	4.13	6.1	Core shifts and cores set wrong.
Housing	B	18.20	20.7	Core shifts and sand holes.
Housing	C	27.42	27.5	Sand holes and core blows.
Housing	C	27.51	28	....
Housing	C	6.43	8.5	Core blows, core shifts, and mis-runs.
Housing	C	2.30	4.3	Cracks and broken.
Carburetor	B	6.44	9	Cracks, sand holes, broken, and core shifts.
Carburetor	C	6.56	8.5	Core shifts.
Fan	C	4.59	6.6	Core shifts, core blows, and bad cores.
Fan	C	3.32	5.3	Core shifts, warped, and cracks.
Hub cap	C	5.03	7	Mis-runs, sand holes, and core shifts.
Cover	C	8.25	10.25	Broken molds and sand holes.
Cap	C	7.12	9	Blows and sand holes.
Union	C	15.14	17	Core shifts, core blows and sand holes.
Average loss		10.54	12.26	

arrived at from an analysis of foundry-production sheets, and the probable losses, obtained by correcting the indicated losses, are tabulated and averaged. The average indicated loss for the kinds of castings for which casting-loss records were available is 10.54 per cent, with a probable loss of 12.26 per cent on the basis of corrected figures. The following figures have also been furnished as to losses in the production of various castings: viz., 1.3 to 10 per cent, with an indicated average of 3 per cent for motor starter and ignition castings; 7 to 18 per cent, with an indicated average of 12 per cent, for a variety of motor castings; 1 to 16 per cent, and an average of 4 per cent, for sand-cast kitchen cooking utensils; and 2 to 20 per cent, with an average of 11 per cent, for vacuum-cleaner castings.

Referring now to the monetary losses involved owing to wasters in castings production, Table 96 gives a summary of the casting losses, together with the conditions affecting losses, and the monetary losses involved at five foundries.

It will be noticed that there is a tendency for the casting losses to be higher with increasing output; this may be explained on the ground that it becomes increasingly difficult to supervise the work of larger numbers of men, although the tendency should be lower in large foundries than in small ones, because the quality of the supervision should be better. Taking the five foundries listed in Table 96 on the basis of a total annual output of 29,220,000 lb. of castings and an average casting loss of 9.4 per cent, the total monetary losses amount to $385,021 per annum. Calculation has been made by the author [27, 35] of the monetary loss incurred in the United States in 1919 because of the scrappage of castings. The melting losses, fuel losses, and machine-shop returns are disregarded; and calculation is made on the basis of an output of 81,000,000 lb. of finished castings and an average loss of 10 per cent for all kinds of castings. Taking the average cost of scrapped castings as $12\frac{1}{2}$ cents per lb., the total monetary loss is $1,125,000, and if the figure 0.8 per cent for machine-shop returns because of defectives be included, then the loss may be placed at approximately $1,200,000. If the average losses were reduced 50 per cent, which is possible by merely eliminating the occurrence of readily avoidable defects, a saving of at least $600,000 would accrue. Material savings are possible in almost every foundry through the elimination of casting losses, and this is a problem that the foundryman may well give detailed consideration.

TABLE 96.—*Monetary losses and other items in representative aluminium-alloy foundries compared.*

Foundry	Total annual amount of metal melted, lb.	Total annual output of finished castings, lb.	Kind of castings made.	Metallurgical control.	Quality of supervision.	Melting and molding equipment.	Quality of records kept.	Average casting loss, per cent.	Annual monetary loss.
A	13,250,000	9,500,000	Crankcases, largely; some other automotive parts.	Fairly good	Good	Modern, good condition	Fairly good	7	$89,383
B	5,400,000	4,000,000	Miscellaneous automotive castings.	Poor	Fair	Modern, good condition	Poor	10	55,555
C	4,500,000	3,000,000	Miscellaneous automotive castings.	Good	Good	Fair	Very poor	8.1	29,086
D	1,420,000	720,000	Vacuum-cleaner castings.	None	Fairly good	Fairly good	Very poor	8	15,652
E	18,000,000	12,000,000	Miscellaneous automotive castings.	Fairly good	Good	Modern, good condition	Good	14	195,345
Totals and averages	42,570,000	29,220,000	....	......	......	......	......	9.4	$385,021

Prevention of Wasters.—While detailed methods for the prevention of specific defects in castings, i.e., blowholes and related defects, and cracks and inclusions, are given in later paragraphs of the present chapter, some general instructions may be found useful. In the first place, accurate scrap-loss records should be kept in which the causes for losses are summarized. Properly kept and interpreted, such records are of the greatest value in diagnosing casting losses; and if a foundryman will study his daily production of defective castings, he will be able to trace the causes for losses, and make such corrections in his practice as will reduce markedly the percentage of wasters. If casting-loss records are to be of any value, they should be kept in a detailed manner, giving the causes for rejection of every casting and the specific nature of the defects. The inspection of castings is an important part in their production, and hasty or inadequate inspection is as useless as inspection by incompetent inspectors. The inspector should preferably be a man familiar with both machine-shop practice and molding, and his decisions should be rendered with impartiality. A competent inspector can readily determine whether a casting is worth machining or whether any surface defect will be removed on machining. He can also be of aid in the reclamation of defective castings by demonstrating how repairs may be made. While the point has not been emphasized hitherto, it should be stated that one of the most effective methods for reducing losses in general is to install an adequate system of mold inspection, whereby every mold is inspected before closing for quality of the workmanship. Many losses due to sand holes, improper setting of cores, wrong gating, and other errors in molding technique can be eliminated by mold inspection.

In preferred practice for handling the inspection of castings, rough inspection is usually made on the foundry floor and after rough cleaning. Any doubtful casting should be placed in a temporary scrap pile for re-inspection. Castings that pass rough inspection are sent to the chipping room for chipping and cleaning, and any castings broken on chipping are also placed in the temporary scrap pile. At the end of a day, the accumulation of defective castings in this pile is re-inspected, and during this scrutiny the production record is made out. All castings that can be salvaged by welding or soldering are placed in one pile, the accepted castings in another pile, and the totally defective ones are returned to the melting room. All accepted castings, including those which have been salvaged by repair, are sent to the sand blast and finally to the machine-shop. Fig. 129 is a diagrammatic outline of the flow of castings through a modern automotive foundry.

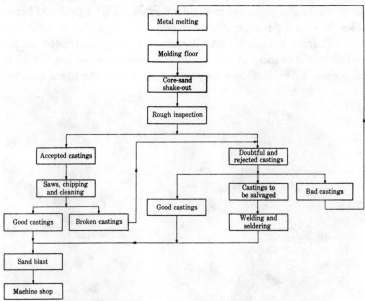

FIG. 129.—*Flow-sheet of castings through an automotive foundry.*

BLOWHOLES, POROSITY, AND UNSOUNDNESS.

Blowholes, porosity, and unsoundness are related defects that are of common occurrence in founding aluminium-alloy castings. It is a fact well known to foundrymen that absolutely sound castings in light aluminium alloys are difficult to produce; and while a casting may be sufficiently sound for some purposes, or while it may be sufficiently free from porosity and related defects to pass certain inspection specifications, it is rarely actually *sound*. An unsound casting must be considered defective in a sense, since any blowhole or porous spot is a source of weakness. Experimental study of the causes for the occurrence of blowholes and related defects in aluminium-alloy castings has been made by the author,[20, 25] and reference may be made to published papers for detailed information.

542 Metallurgy of Aluminium and Aluminium Alloys

Blowholes may be regarded as spaces, in a frozen alloy, filled with air or other entrapped gas which is the primary cause of the blowholes. Large holes are not usual in sand castings, and holes 0.025 to 0.075 inch in diameter and microscopic holes are the rule. Fairly large blowholes up to $\frac{1}{4}$ inch in diameter are common in die castings and permanent-mold castings. Fig. 130 shows blowholes in 91 : 8 : 1 aluminium-copper-iron alloy, taken on a section cut from a boss on an oil pan. Porosity is defined as the inability of a casting to withstand pressure, or a casting is said to be porous when it shows seepage leaks under the open gasoline

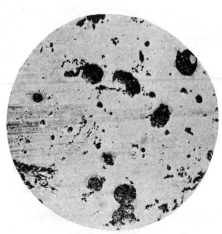

FIG. 130.—*Blowholes in a section cut from a boss on an oil pan; unetched; vertical illumination;* $\times$ 5.

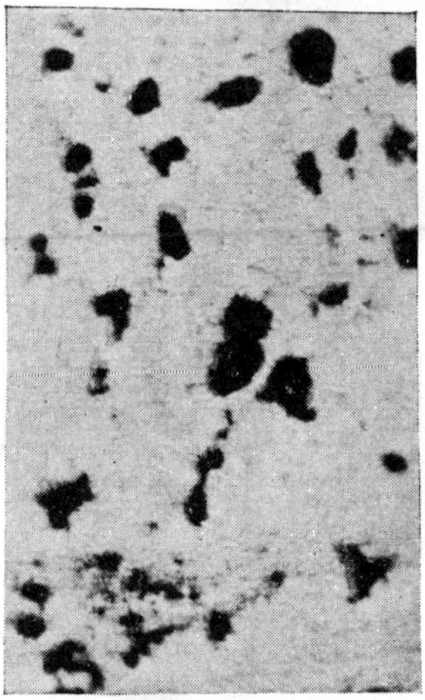

FIG. 131.—*Porosity in a section cut from an oil pan; unetched; vertical illumination;* $\times$ 7.

test. Porosity may be due to blowholes, cracks, pinholes, general unsoundness, occluded foreign matter, or to a combination of these defects. Fig. 131 is a macrograph of a section from a casting which showed seepage leaks under the open gasoline test. Unsoundness is an undesirable term at best; it is used indiscriminately to connote that condition in a casting where flaws exist due to blowholes, draws, small cracks, pinholes, and occluded foreign matter.

Factors Affecting Blowholes and Related Defects.—In any casting, the occurrence of blowholes, porosity, and unsoundness is due to a number of factors, of which the following are the most important: (1) high melting and pouring temperatures; (2) dissolution of gases; (3) constitution of the melting-furnace atmosphere; (4) method of molding; (5) design of the castings; (6) composition of the alloy, and (7) quality of the melting stock. The effects of these and other factors on the occurrence of blowholes and related defects in aluminium-alloy castings have been discussed at length by the author in another place,[20] and can be treated here only very briefly.

As is well known, blowholes in metals and alloys are intimately associated with gases, and blowholes, porosity, and unsoundness are ascribed to gases dissolved in the liquid alloy, which are liberated on final freezing. According to Henry's law, liquids, in general, dissolve less gas with rising temperature, but liquid metals depart from this law and dissolve more gas with rising temperature.[7] Moreover, such gas-metal solutions readily remain saturated; consequently, the higher the melting temperature the more gas will be dissolved, and the more blowholes in the resultant castings. Then too, the pouring temperature is a cogent factor in influencing the soundness of castings, and, in general, the higher the pouring temperature the greater the unsoundness. Cooling from a high melting temperature to a low pouring temperature will help overcome the deleterious effects of the former, but any casting that has been poured from a highly overheated melt will be more unsound than one poured from a melt not overheated, both being cast at a low temperature. The method of melting, or more precisely, the atmosphere of the melting furnace, is an important factor. In an oxidizing atmosphere, there will be ample opportunity for aluminium oxide to be formed and for nitrogen and oxygen to be taken up, and, in general, castings poured from alloys melted under oxidizing conditions may be expected to be relatively unsound. The method of molding is an item of importance for consideration in the study of unsoundness, and blowholes may be caused by gases set free by the molds or cores. Liquid metal can not remain in an impervious mold because the gases evolved must find an outlet, and will escape by throwing the metal through any possible opening, causing a "kick-back" through a runner or ejection through a riser. The gases evolved on pouring liquid alloys must be drawn through the sand, and it is necessary usually to augment the porosity of the sand by venting.

The effect of castings design is an important factor affecting the occurrence of blowholes, and the design and other factors affect the rate of solidification. In pouring a casting having angles and corners, the alloy in these places is the last to solidify; hence, in addition to shrinkage cracks forming therein, there is a tendency for gases and other impurities to be forced into them, causing local unsoundness. Thick and thin sections in contiguity should be avoided, as well as designs that give rise to columnar crystallization. The growth of columnar crystals brings about planes of weakness, and unsoundness is frequently an accompaniment of such a condition. Thus, Fig. 132 shows the macrostructure of a boss section cut from a cracked crankcase in 92 : 8 aluminium-copper alloy; the two sets of columnar crystals meeting in a line may be observed, as well as the unsoundness. Microscopic examination of many samples of cast aluminium alloys [32] has shown that unsoundness and weakness is generally present (1) at the juncture of columnar and equiaxed crystals; (2) at the juncture of two sets of columnar crystals; and (3) at the juncture of any two columnar crystals. This would be expected from theoretical considerations. While substantiated experimental observations as to the effect of the chemical composition of the alloy upon soundness have not been made, it is well known that some light aluminium alloys tend to be more unsound on casting than others. Thus, the tin-bearing aluminium-copper alloys are more sound than the aluminium-copper alloys without tin, and the silicon-bearing aluminium alloys are more sound than the usual casting alloys. Much uncertainty exists as to the effect of the quality of aluminium pig on the soundness of the resultant castings, but tests have shown that pig which yields low values in the tensile test, and which will not bend much in the bend test, normally yields poor castings,

FIG. 132.—*Thick and thin columnar crystals in 92 : 8 aluminium-copper alloy; etched NaOH; oblique illumination;* × *1.5.*

which may be excessively hot-short, very porous, and unduly cracked.

Author's Experiments on Blowholes.—Experimental study has been made by the present author of the question of blowholes, porosity, and unsoundness, and the results reported.[20, 25] Tests were made as to the effect of (1) pouring temperature, and (2) cooling to various pouring temperatures after heating to higher melting temperatures, on the occurrence of blowholes and related defects in 92 : 8 aluminium-copper

FIG. 133.—*Unsoundness and blowholes in 92 : 8 aluminium-copper alloy, poured at 640° C.; lightly etched NaOH; vertical illumination;* ×7.

FIG. 134.—*Unsoundness and blowholes in 92 : 8 aluminium-copper alloy, poured at 950° C.; lightly etched NaOH; vertical illumination;* × 7.

alloy sand castings. The castings were examined microscopically and macroscopically for unsoundness. Fig. 133 shows the surface appearance of a section cut from a casting poured at 640° C., and Fig. 134 shows a similar section from a casting poured at 950° C. These macrographs indicate the effect of pouring temperature, the number and size of blowholes being found to increase with increasing pouring temperature. Microscopic and macroscopic examination of the castings gave evidence, also, of at least three different kinds of unsoundness, viz., (1) that which is plainly intergranular and may be due to an actual forcing apart of the grains by gas attempting to escape, or due to intergranular occluded foreign matter; (2) that due to liberation of gas which is entrapped at the moment of final freezing and can not escape, resulting in blowholes of varying size; (3) that which is the result of "balled-up" aluminium oxide occluded indiscriminately in rather large "gobs" in the frozen alloy. Fig. 135 shows intergranular occluded matter, whether aluminium oxide or not, in a casting.

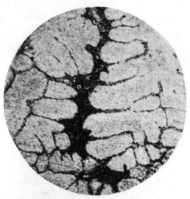

FIG. 135.—*Intergranular unsoundness in 92 : 8 aluminium-copper alloy; poured at 850° C.; etched NaOH;* × 75.

While the question of cracks in aluminium-alloy castings will be discussed in a later sub-division of this chapter, it is of importance to state here that cracking may be associated with blowholes, and in the author's investigation on unsoundness it was shown by fracture studies that some surface cracks are intimately associated with deep-seated internal blowholes. In a number of sections, it was found that surface cracks penetrated into an interior blowhole of rather large size, say $\frac{1}{4}$ in. diameter, and that the blowhole evidently caused the crack because of attempt of entrapped gas to escape. Thus, Fig. 136 shows the macrostructure of a section cut from a badly cracked crankcase from a portion in the vicinity of surface cracks. In examining a fracture through

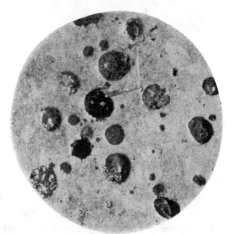

FIG. 136.—*Blowholes in a sample taken from a crankcase;* × 5.

the section, internal blowholes, shown at point X of Fig. 137, were found, from which cracks led to the surface.

The main conclusions arrived at from the experimental work may be summarized as follows:

1. The number of blowholes present is a function of the pouring temperature; the higher the pouring temperature the greater the number of blowholes and the more unsound is the casting.

2. Unsoundness varies with the temperature to which the charge is heated; the higher the temperature in the furnace the more unsound the resultant castings are, irrespective of the pouring temperature.

3. Unsoundness is a function of the length of time of melting; the longer any melt is held in the furnace the more unsound are the castings, irrespective of the temperature of heating and the pouring temperature.

Methods for Preventing Blowholes.—It might be inferred that the author believes that entirely sound castings can be made at all times, but no such inference should be drawn. Evidently,

Fig. 137.—*Internal hole shown on fracture in a sample removed from a crankcase;* $\times 1$.

however, many castings are needlessly unsound. While it is not possible to give a code of directions which a melter can follow and thereby obtain 100 per cent sound castings, this much is well settled, viz., that it is possible to control the variables that may conduce to unsoundness and blowholes, and thereby eliminate much of the usual difficulty. Pouring an overheated melt at a lower temperature by allowing the charge to cool prior to pouring will aid in minimizing the deleterious effects of overheating. Castings poured at low temperatures are more sound than those poured at high temperatures, but heats held in the furnace for a long time at either high or low temperatures are more unsound than those held for a short time. The most aggravated cases of unsoundness will result from pouring a melt at a high temperature that has been previously excessively overheated and for a long time. There is some evidence for the belief that there is a minimum temperature below which it is not safe to go for fear of unsound castings resulting from too low pouring temperature. In general, however, with the methods of melting now in vogue, the heats should be kept at a low temperature in the furnace, melting should be as rapid as possible, the charge should be poured as soon as it is melted, and the pouring temperature should be as low as is consistent with the metal filling the mold.

Methods of Testing for Porosity.—In practice, it is necessary on inspection to test certain types of castings for porosity and related defects, and the usual methods applied are discussed briefly below.

The usual test employed for detecting general or local porosity, in such castings as crankcases, oil pans, manifolds, etc., is the open test with a solution of methylene blue in gasoline. The test is carried out by painting the liquid, or smearing with a rag swab, on one side of the casting at points where porosity is suspected or might be dangerous. If the casting is porous, the solution will seep through and show on the other side. For some hollow castings, such as carburetor bodies, the casting may be filled with the solution and seepage watched for. The main advantage of the test lies in its simplicity, and it is the most generally used test in foundry practice. Ordinarily, on application of the solution, seepage leaks may be classed as small spots not more than $\frac{3}{8}$ inch in diameter, which show, as indicated by the appearance of the blue color, on the opposite side of the casting, after about 10 mins. Bad porosity will show up in many cases by the almost immediate seepage of the solution through the casting. Other liquids may be employed instead of gasoline, e.g., xylol.

Another test for porosity is the metallographic test used by the author.[32] In carrying out the test, samples are cut from the gates or risers, and polished as in metallography. Ocular examination or inspection under a low-power lens will give an idea of the defects present. Broadly speaking, if blowholes and porosity are found in the gates and runners, they will also be found in the actual casting. The presence of minute porosity in aluminium-alloy castings can be detected, as suggested by Rawdon,[33] by immersing a polished section in alcohol colored with picric acid or some other bright colored dye. The section, after being dried, is allowed to stand in a warm place. Porosity is then indicated by the appearance of colored spots on the surface, owing to evaporation of the colored alcohol from the internal cavities where it was enclosed.

The third type of porosity test is the pressure test, and there are three general pressure tests which are applied to aluminium-alloy sand castings for the detection of porosity and leaks, viz., (1) the water-pressure test; (2) the air-pressure test; and (3) the steam-pressure test. The air-pressure test is rather more largely employed of the three, but the other two are distinctly useful for specific purposes. The water-pressure test is carried out simply by leading water under pressure into the closed casting and noting the leakage, if any, through the walls. The water pressure is variable, depending upon requirements, but 30 to 40 lb. is ordinarily used. In the case of the air-pressure test the casting, e.g., carburetor bodies, is connected to an air line, after being suitably plugged, and then immersed in a tank of water. Air pressure is applied, and any leaks will be readily detected by air bubbles issuing from the walls of the casting. Pressure may be applied up to 100 lb. per sq. in., more or less, depending upon what pressure the particular casting will withstand. As to the steam-pressure test, this is useful for many kinds of castings, but it is generally applied only to castings which must withstand steam pressure in service. This test is applied by simply introducing steam under a pressure of 50 to 100 lb. into the casting and noting any leaks.

A method of testing for porosity, which furnishes numerical data, has been devised by Bezzenberger and Wilkins.[31] The advantage of such a test would obviously be great when it is desired to make comparison of different alloys cast under identical conditions. In carrying out this test, a cast test cup of the form shown at the lower right-hand corner of Fig. 138 is used. This cup is cast either in sand or in a chill mold, depending upon the tests to be made, and it may be cast either oversize and then machined to size, or else cast to size with the skin on. The testing device is shown at the left in Fig. 138, with the cup in place. The cup is placed on a rubber gasket resting on the base A, which is made of steel and drilled so that air may be admitted into the test cup. B is a steel flange which is screwed down on the base A, thus holding the test cup in place and giving an air-tight joint between the flange of the cup, the rubber gasket, and the base. C is a small copper tube with a connection of rubber tubing at one end. A 2,000-c.c. glass graduate is inverted over the test cup and tube C, as shown in the figure, and the whole apparatus is placed in a small tank partly filled with water.

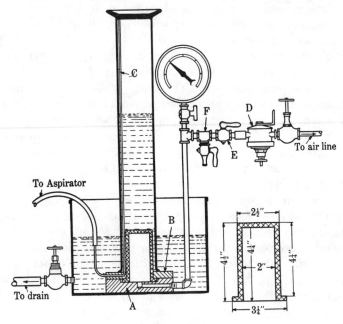

Fig. 138.—*Apparatus and test cup for determining the porosity of castings (Bezzenberger and Wilkins).*

Tube C is connected to an aspirator which is employed to draw off air from the glass graduate. Any air so removed is replaced, of course, by water. Connections are made, as shown in Fig. 138, to an air line under constant pressure. D is a reducing valve used for maintaining any desired air pressure. E is a cock for the admission of air to the apparatus proper, and F is another cock which opens the system to the atmospheric air. Cock F is used to release the pressure immediately upon the completion of a test. A pressure gage is inserted in the air line, as shown, and a thermometer is placed in the tank of water.

Casting Losses and Defects in Castings

In carrying out a test for porosity upon a cast cup, the cup is placed in position resting on the rubber gasket on the base A. The flange B is then screwed down so as to give a tight joint, and the water tank is then filled until the water is at a level about half-way from the top of the test cup as placed in position. Cock E is now opened for the purpose of testing the apparatus for leaks, and a reading of the pressure gage is taken. When required, the reducing valve D is adjusted to give the desired pressure. The air is then shut off, and cock F is opened until the pressure is reduced to atmospheric pressure; it is then closed. The glass graduate is then placed over the cup and the tube C is filled with water by means of the apirator connected to the tube. The set-up for the test is now in readiness. Cock E is opened at an observed time, and any air which leaks through the cast cup is collected in the glass graduate. When the volume of air collected is sufficient so that a good reading may be taken, the cock E is closed, cock F is opened, and the time taken. These last three operations should be made at once. Usually, about 500 to 1,000 c.c. of air should be allowed to collect in the graduate; a larger volume is undesirable because the relative error is in inverse proportion to the volume. The difference in level between the water in the graduate and that in the tank is measured, and the temperature of the water is taken. The test is now completed, and the following data should have been taken: (1) volume of air collected; (2) pressure of air employed; (3) difference in water levels in the graduate and in the tank; (4) time of passage of the air; (5) temperature of the water; and (6) the barometric pressure.

The porosity of the cup under test may then be calculated from the formula

$$K = V \times \frac{273}{Ta} \times \frac{B - Vp - \frac{H}{13.6}}{760} \times \frac{1}{Tm \times Pa}$$

where K = the measure of porosity,
 V = volume of air collected, in c.c.,
 Ta = absolute temperature of the water, in ° C.,
 B = barometric pressure, in mm. of mercury,
 Vp = vapor tension of water at the observed temperature, in mm. of mercury,
 H = difference in water levels, in mm.,
 Tm = time of passage of air, in minutes, and
 Pa = pressure used, in atmospheres.

The measure of porosity, K, is numerically equal to the number of cubic centimeters of air, under standard conditions, leaking through the test cup per min. per atmosphere of pressure. The term Pa in the formula is introduced for the purpose of simplifying the number K. For comparative results, Pa should be constant in individual tests. The proportion,

$$\frac{V}{V_1} = \frac{Pa}{Pa_1}$$

while approximate, is not exact. The test cup can be altered as to design so as to permit the study of sections of various thicknesses and of different areas for porosity.

Treating Porous Castings.—There are a number of methods used for the treatment of porous castings, designed to fill the holes and unsound portions so that the castings will pass inspection tests. Some of the methods which have either been suggested or used are entirely worthless, if not actually harmful to the castings, and these should not be permitted in the treatment of automotive or other important parts.

The usual method employed for treating porosity is the sodium-silicate method. In this, the castings are first soaked in concentrated (40° Bé., 1.38 sp. gr.) sodium-silicate solution for at least one hour. The solution may be made up by adding about four times its own volume of water to the sodium silicate. The time period of soaking varies with different size castings and also with the thickness and porosity. The soaking treatment is followed by quick dip (1 to 2 mins. immersion) in a dilute acid such as 25 per cent sulphuric acid, 5 per cent hydrochloric acid, or 10 per cent sodium acid sulphate. Following the acid dip, the casting is washed in cold water to remove the acid and soluble salts, and then dried at above 100° C. for at least one hour. The time period of heating after washing depends, of course, upon the size and thickness of the casting, since the silica from the sodium silicate must be at least partly dehydrated by the heating to render it unaffected by ordinary weathering conditions. In British practice for the treatment of porous aviation-engine castings, such as inlet and exhaust manifolds, water pipes, radiator castings, and the like, the parts are treated by pumping a hot solution of sodium silicate at 40 to 60° C. under 70 to 200 lb. pressure into the plugged castings. The pressure is maintained until sweating stops, and with very porous castings, holding 70 lb. pressure for 20 mins. is ordinarily ample. With 200 lb. pressure, about 10 mins. will ordinarily suffice for very porous castings, while with the majority of castings holding this pressure for 3 to 4 mins. will stop any leaks. When the pressure has been held sufficiently long, it is released, the solution poured out, and the casting is washed thoroughly with hot water. It is important that the casting should not be allowed to dry before washing and thus allow the solution to fall into powder, but immediately after running out the solution the casting must be rinsed with hot water. Any white powder that has deposited on the surface should be brushed off with a wire scratch brush, and the casting is then allowed to stand until thoroughly dry. Treatment with sodium-silicate solution may be given either before or after machining. Botta [42] has discussed the impregnation of aluminium-alloy castings with sodium silicate. While commercial sodium silicate, $Na_2Si_2O_5$, (water glass) is ordinarily used, apparently better results are obtained with sodium metasilicate, Na_2SiO_3, since the alkalinity is greater. The specification given by the Bureau of Aeronautics, U. S. Navy Dept., for sodium silicate to be used for castings treatment is as follows:

>Alkalinity (calculated as NaO)..............14.0 per cent.
>Water, not more than.....................45.0 per cent.
>Sp. gr. at 65° F., not less than..............1.7.

Other methods employed in treating for porosity include the following: (1) soaking in concentrated solution of zinc chloride, followed by washing; (2) soaking in a solution of aluminium chloride, followed by dipping in ammonia water, and heating above 100° C.; (3) soaking in a sodium acetate-ammonium chloride solution, followed by heating above 100° C.; (4) treatment with bakelite under pressure; (5) treatment with bakelite and aluminium powder; (6) boiling in linseed oil under pressure, followed by baking; (7) annealing at 300° C.; (8) doping with aluminium powder and sulphur; and (9) soldering or welding. The first three soaking methods are similar in principle to the sodium-silicate treatment and are efficacious. In using bakelite, the liquid may be forced into the castings under pressure, or the castings may be simply dipped into the substance. The material known as Bakelite No. 6 is recommended. In the case of hollow castings, the enamel is poured in and held until it has penetrated the porous spots, and the excess material poured out. After treatment with bakelite, the casting should be air-dried for a short time, and then baked for two hours at 150° C. With large castings, where a few porous spots have been detected by the open gasoline test, the enamel may be applied with a hand brush or an air spray. Several coats should be applied if the bakelite is used in this way. When the spraying process is used, care should be taken to keep the filler of this enamel in suspension. The de Vilbiss sprayer with an air agitator is suitable for this purpose. It will assist the penetration of the enamel, if the casting is heated in the vicinity of the porous spots to about 100° C. before application. Bakelite is at times mixed with aluminium powder when applied. Another method of treatment consists in pumping boiled linseed oil under pressure into the casting, following by baking at moderate temperature. Annealing at 300° C. has been suggested and actually used at some foundries in the treatment of porosity, but it is difficult to discern what the effect can possibly be. The use of a mixture of aluminium powder and sulphur has been suggested and used for doping porous spots, but the author can not too strongly condemn this method. It is of no value, if not actually harmful, and there can be no possible object in using it, from the metallurgical point of view. Porous spots are sometimes closed by peening, i.e., hammering with the ball peen of a machinist's hammer, but this method is not reliable. Porous and unsound places in castings are, at times, treated by soldering and welding, and application of soldering and welding to casting salvage has been discussed in detail by the present author and M. E. Boyd.[47]

INCLUSIONS IN ALUMINIUM-ALLOY CASTINGS.

The occurrence of hard inclusions, i.e., the so-called hard spots of foundry parlance, in aluminium-alloy castings, which give rise to difficulties in machining and polishing, is of interest and importance to founders and users of parts. In the founding of certain castings, the percentage of machine-shop returns owing to rejection because of hard inclusions may be high, and be the cause of considerable loss. The defect is troublesome in both sand castings and die castings. Hard inclusions are so widely different in character that the term hard spots is only roughly descriptive. Hard spots are ordinarily defined as any kind of metallic or non-metallic inclusions that cause difficulty on polishing or machining. Ordinarily, hard inclusions are not noticed during the routine inspection of rough castings in the foundry, but they are found during machining operations. When a machine tool strikes a hard inclusion, its edges are quickly

FIG. 139.—*Piece of brick in gate from a crankcase;* × 1.

dulled and rendered unfit for cutting, and breakage may at times occur. Investigation of the causes for, and methods of, preventing hard inclusions has been made by the author,[26, 38] and reference may be made to this work for detailed information.

Kinds and Characteristics of Hard Inclusions.—All hard inclusions occurring in aluminium-alloy castings may be divided

FIG. 140.—*Piece of stone in gate from vacuum-cleaner nozzle;* × 2.

into two classes, viz., non-metallic inclusions, and metallic inclusions. Under the former are included all hard, foreign, non-metallic inclusions such as pieces of brick or cement, chunks of crucibles, core sand and molding sand, and hard clay. Non-metallic inclusions range in size from microscopic particles to chunks $\frac{1}{2}$-in. cube or larger. Typical examples of non-metallic hard inclusions are shown in the accompanying illustrations.

556 Metallurgy of Aluminium and Aluminium Alloys

Thus, Fig. 139 shows a piece of brick caught in a gate from a crankcase, and Fig. 140 shows a piece of stone caught in a vacuum-cleaner nozzle gate. Large inclusions of the type shown

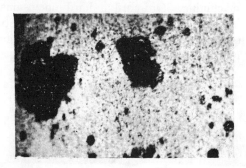

FIG. 141.—*Stony hard spots in vacuum-cleaner fan; unetched; vertical illumination;* × 7.

are not likely to be numerous, but small inclusions tend to be numerous, when they are present at all. Fig. 141 shows some fairly large stony inclusions in a vacuum-cleaner fan, while Fig. 142 shows some crucible bits in the same sample.

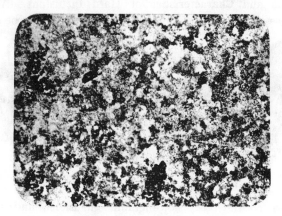

FIG. 142.—*Inclusions of bits of crucible in vacuum-cleaner fan; etched NaOH; oblique illumination;* × 7.

Roughly speaking, so-called metallic hard inclusions are generally traceable to iron. Thus, the iron may be so high in No. 12 alloy that large crystals of the intermetallic compound $FeAl_3$ will be formed; this compound is hard and brittle, and when the iron exceeds 2 per cent, difficulty may be experienced on machining and usually will be on polishing. So-called " inclusions " of the iron-aluminium compound FeAl₃ are not really inclusions, but the presence of needles of FeAl₃ in No. 12 alloy is due simply to the fact that the compound is insoluble in the alloy. Actual irony spots (true inclusions) may arise from charging nails, core wire, iron chaplets, chills, or other bits and pieces of iron into the furnace. Fig. 143 shows crystals of FeAl₃ in an aluminium-alloy

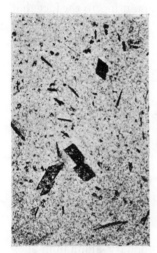

FIG. 143.—*Crystals of FeAl₃ in a die casting that polished poorly; etched NaOH; oblique illumination;* × 8.

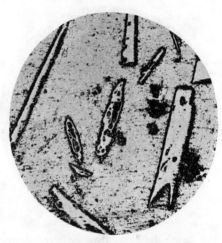

FIG. 144.—*Crystals of FeAl₃ in unetched section from same sample as in Fig. 143;* × 30.

die casting, high in iron, which gave difficulty on machining and polishing. Fig. 144 is a photomicrograph of the same section. When an aluminium alloy high in iron is polished, the softer matrix is worn away rapidly, but the harder FeAl₃ is not readily worn down and consequently stands in relief, giving a rough appearance to the surface. Fig. 100 (Chapter IX) shows large inclusions of brittle iron-pot scale included in 92 : 8 aluminium-copper alloy, and Fig. 145 shows the microstructure of the large inclusion in Fig. 100. The white areas in Fig. 145 correspond to the white specks in the large inclusion in Fig. 100; these areas are FeAl₃.

Factors Affecting the Occurrence of Inclusions.—The essential cause of actual non-metallic foreign inclusions in aluminium-alloy castings is the charging of foreign materials, e.g., pieces of brick or stone, pieces of crucibles, core sand and molding sand, and other extraneous substances, to the melting furnaces. Such materials are numerous in foundry-floor sweepings, and non-metallic inclusions are more likely to be present when all-scrap or high-scrap charges are run than when all-primary melting stock is used. At the same time, there is substantiated evidence pointing to the fact that such inclusions may be caused by primary melting stock, and a case in point will be cited in a later paragraph. New iron pots may sometimes cause hard inclusions because of the sand and scale adhering to them. Hard inclusions caused by iron may be due to four main causes: (1) the use of secondary aluminium pig and secondary aluminium-alloy pig, high in iron; (2) the charging of pieces of iron, such as nails, chills, core wire, steel chips and drillings, and the like with the foundry scrap; (3) the dissolution of iron from the pots in iron-pot melting, and (4) the hard, brittle complex iron-aluminium alloy that forms on the sides of iron pots.

FIG. 145.—*Microstructure of the large inclusion in Fig. 100 (Chapter IX); $FeAl_3$, white; unetched;* $\times$ *100.*

Among other factors that may be mentioned as affecting the occurrence of hard inclusions in castings are the following: (1) size of the castings; (2) machining and polishing; (3) method of molding; (4) method of melting; and (5) quality of the melting stock. These factors have been discussed at length elsewhere,[26, 38] and need be considered only briefly here. Hard inclusions are presumed to be more numerous in small castings than large ones, but there is no basis in fact for this belief. The apparent predominance of inclusions in small castings may be explained on the ground that these are more generally polished. As indicated previously, hard inclusions are seldom found in the routine foundry inspection of castings, but they are revealed during machining or polishing. Hence, if a given type of casting requires much machining, complaints from the machine-shop regarding hard inclusions are likely to be more numerous than when castings require little machining. The method of molding is a factor governing the occurrence of hard inclusions to a subordinate extent. Skim gates have been used in the attempt to eliminate this defect, and while they may be of some value in barring out large included particles, they will not keep small particles out of the castings. Shrink balls and blind risers have also been suggested and used to prevent hard inclusions, but these devices are quite useless for the purpose. The gating of castings is important, and influences the occurrence of hard inclusions. If a sharp constriction be placed in a runner leading from a main gate to the casting, larger foreign particles may be trapped in the constricted area and prevented from entering the casting. On the other hand, if such a constriction is at the point where the runner joins the casting, the trapped inclusions will be frozen partly in the runner and partly in the casting; then, if machining must be done at this point on the casting, hard spots will be encountered. An examination of the method of gating and molding will assist a foundryman to determine whether any feature here may be corrected.

The effect of the method of melting has been alluded to previously. Iron-pot melting presents some disadvantages which other methods do not have, when considered in relation to the occurrence of hard inclusions, since iron may be dissolved and chunks of the hard ferro-aluminium alloy (iron-pot scale) may be knocked off into the melt. In crucible melting, pieces and chunks may be knocked off old and friable crucibles on lifting, stirring, and pouring. The effect of the quality of the melting stock has also been alluded to. Castings poured from primary melting materials are less likely to contain foreign included matter or crystals of $FeAl_3$ than such castings poured from charges containing much scrap. It has also been pointed out that hard non-metallic inclusions found in castings have been traced to the use of certain primary melting stock, and it is of interest to cite an actual case in this connection. In the production of automotive crankcases at a certain foundry, it was found that when a certain brand of primary aluminium pig was employed for making up No. 12 alloy, a peculiar form of non-metallic hard spots occurred in the resultant castings. When a different brand of aluminium pig was employed, all other conditions being the same, the hard inclusions disappeared and were no longer found in the castings. It is also of interest to point out that when the brand of aluminium which caused the difficulty was melted, either in open-flame or iron-pot furnaces, large hard lumps were formed in the melt, similar to those hard inclusions found in the castings. The actual composition of one of the lumps formed on melting was as follows: copper, 2.92 per cent; iron, 0.51; alumina (soluble), 14.75; silica (including insoluble alumina plus siliceous matter), 34.64; and aluminium (by difference), 45.16 per cent. The original pig was substantially 99 per cent grade metal, but was found to be exceptionally dirty on microscopic examination; it apparently contained considerable alumina or included bath salts from the reduction cell or other foreign impurities. The chemical analysis of the pig did not indicate anything to which the hard lumps might be attributed, but it was thought that its oxidized condition might have been responsible for the formation of nuclei upon which an aggregation of impurities might have grown. Fig. 146 shows a piece cut from a crankcase containing a large hard lump, which was revealed on machining. Other cases similar to this one have been observed, and there seems to be no question but that hard non-metallic inclusions can be traced at times to the quality of the primary aluminium.

FIG. 146.—*Hard non-metallic inclusion in crankcase; about ⅔ actual size.*

Author's Experiments on Hard Inclusions.—In studying the causes for and prevention of inclusions in aluminium-alloy castings, experiments were made by the author,[26, 38] to produce inclusions in 92 : 8 aluminium-copper alloy at will, with the view to ascertaining the conditions that might influence the occurrence of this defect. A set of vacuum-cleaner fans—a type of casting in which hard inclusions are prone to occur—was poured under various conditions, and a summary of the results of these experiments is given in Table 97. The details of these experiments need not be given here, but the principal conclusions arrived at may be indicated. It may be pointed out in passing that a convenient foundry method for detecting the presence of hard spots before machining is to examine fractures of the gates. This procedure will avoid any unnecessary spoiling of the castings. If the lead-off gate or runner from the main gate to the individul castings is constricted in area at any point, numerous inclusions (which may be hard) will be trapped at the point of constriction. Hence, when the gates are fractured or the castings simply broken off from their runners, foreign included matter may be seen in the fractures. If the hard inclusions are microscopic, they will not be revealed at the sprue cutter or the band saw, but if large, they may readily be so revealed. In the experiments mentioned here, the fans were broken off from the lead-off gates, and the fractures were examined for foreign inclusions visible to the eye.

Reverting to the experimental data summarized in Table 97, there were six castings poured in each of the experiments A to F, and three each were machined according to the regular operations called for by the job. All of the castings were machineable, but they varied somewhat in machining qualities. According to the appearance of the machined surfaces, the castings were rated as to machineability and freedom from hard inclusions in the following order, D, C, F, E, B, A. According to this rating, all-primary materials melted in graphite-clay crucibles gave better results than the other melts used. While the observation is not entirely conclusive, it might be expected that crucible melting, when prime materials were used, would yield better castings than iron-pot melting when all-scrap or part-scrap charging materials were used. This is further corroborated by examination of polished microsections cut from the main gates in the respective experiments. Three of each set of fans cast in experiments A to F, inclusive were polished and buffed according to usual shop practice. From visual observation the castings

562   Metallurgy of Aluminium and Aluminium Alloys

TABLE 97.—*Summary of experiments made to examine the effect of some factors on the occurrence of hard inclusions in castings.*

Experiment.	Method of melting.	Composition of charge.	Skimming.	Metal taken from.	Flux used.	Remarks as to appearance of fractures; six fractures in each case.	Rating on the basis of appearance of machined surfaces.	Rating on the basis of appearance of polished surfaces.
A	Iron pot	All gates, risers, and defective castings.	None	Top of pot	None	Many large and small inclusions in every fracture.	6	6
B	Iron pot	All gates, risers, and defective castings.	Well skimmed	Middle of pot	None	Large inclusions in 2 fractures; 4 fractures clean, much better appearance than A.	5	5
C	Plumbago crucible	All primary aluminium pig plus 50 : 50 Cu-Al alloy.	Well skimmed	Top of crucible	None	Small inclusions in 5 fractures; 1 fracture clean.	2	2
D	Plumbago crucible	All primary aluminium pig plus 50 : 50 Cu-Al alloy.	Well skimmed	Bottom of crucible	None	Small inclusions in 2 fractures; 4 fractures clean.	1	1
E	Iron pot	50 : 50 per cent gates and defective castings plus 50 per cent primary aluminium pig and 50 : 50 Cu-Al alloy.	Well skimmed	Top of pot	Sal ammoniac	Large and small inclusions in 3 fractures; 3 fractures clean.	4	3
F	Iron pot	50 per cent gates and defective castings plus 50 per cent primary aluminium pig and 50 : 50 Cu-Al alloy.	Well skimmed	Bottom of pot	None	Large and small inclusions in 3 fractures; 3 fractures clean.	3	4

were rated on appearance as follows: D, C, E, F, B, A. In other words, fewer noticeable inclusions were revealed on polishing the faces and less dragging and scratching of the polished surface by hard occluded particles was noted in the order given. Thus, with reference to the appearance of the polished surfaces, the castings are rated in practically the same order as those examined after machining.

Iron-Pot Scale.—As mentioned, hard inclusions due to iron-pot melting may be traced at times to the hard complex ferro-aluminium alloy that forms on the sides of the pots. If aluminium or one of its light alloys is kept liquid in contact with cast iron, there is a gradual dissolution of iron in the aluminium, and the concomitant formation of an iron-aluminium alloy on the sides of the pot. The iron content of this alloy varies throughout the section of the alloy adhering to the pot, being richest in iron in that part of the alloy adjacent to the pot and leaner in iron nearer the liquid aluminium. If liquid aluminium is kept in undisturbed contact with a cast-iron pot for a long time, the scale gradually builds up in thickness to $\frac{1}{2}$ inch or more on the sides and bottom of the pot. If a considerable thickness of this scale builds up, it becomes very adherent and is not easily dislodged, but small accumulations are readily knocked off. Eight samples of iron-pot scale had the following range of composition: iron, 1.39 to 13.40 per cent; silicon, 0.41 to 0.47; total carbon, up to 2.15; and aluminium, 86.40 to 98.20 per cent (by difference). The iron content of scale scraped from the inside of melting pots is very variable, and the scale itself is a complex alloy containing iron, aluminium, silicon, carbon and other elements in subordinate amount. Fig. 99 (Chapter IX) shows the fracture of a piece of hard iron-pot scale taken from a cast-iron melting pot.

Composition of Hard Inclusions.—As indicated, the chemical composition of hard inclusions in castings may be very variable, depending upon the type of the inclusion, and visual examination will usually suffice to determine their character. If inclusions are very small, microscopic examination is necessary to determine their character. The chemical composition of hard inclusions may be determined only with much difficulty, because it is almost impossible to separate such inclusions cleanly from the castings. Chemical analyses of large inclusions have shown them to be largely silica (chunks of stones), iron-pot scale, and pieces of graphite-clay crucibles. Several samples of small hard inclusions have been isolated and analyzed, but they were wetted with aluminium and consequently run high in that element. In certain samples, silicon was 0 to 6 per cent, aluminium 59 to 87 per cent, and iron in subordinate amount. Lyon [44] has recently investigated the cause of hard spots in aluminium-alloy castings, finding that irony spots are oxides of iron, which probably come from the hardener used to introduce iron into the alloy or from the scale on stirring rods or iron pots. Lobley [45] has discussed the question of non-metallic inclusions in aluminium, pointing out that the reduction-cell product may contain some adsorbed cryolite with or without alumina in solution.

Methods for Preventing Hard Inclusions.—Careful attention to certain details of foundry practice will go a long way in preventing the occurrence of hard inclusions which cause difficulty on machining, and some suggested methods of prevention follow:

As explained, inclusions originate largely from dirty melting practice, and care should be taken to keep the melting charges clean. When the melting room and foundry floor are being cleaned (to gather up small pieces of metal, sloppings, over-runs, and the like) dirt, gravel, sand, cement, pieces of brick, and chunks of broken crucibles, etc., are swept up. Floor sweepings should be screened before charging, and small scrap should be forked. In crucible melting, particularly where the crucibles are old, badly cracked, and weak, pieces of the crucible may fall or be knocked into the metal and carried into the castings. Where graphite-clay crucibles are used for pouring with any kind of melting practice, pieces of the crucible may be found in the castings; hence crucibles should not be used too long or until they become cracked and readily friable. No core sand should be left in defective castings that are re-melted, because the hard, fine silica particles may give rise to small hard inclusions. Foreign non-metallic inclusions can be largely eliminated, provided sufficient attention is given to the charging practice.

The status of conditions causing hard inclusions owing directly or indirectly to iron has been explained. The iron content of No. 12 alloy or other alloys made from secondary pig can be kept within reasonable limits by using only such pig as is reasonably low in iron. Difficulties brought about by the use of secondary pig, or scrap castings purchased from outside sources, can be readily eliminated by chemical analytical control. Actual irony hard spots caused by charging nails, core wire, chills, and bits of iron and steel can be prevented by careful attention to the quality of the charge. Mechanically admixed iron appearing in the foundry scrap or in floor sweepings can be removed by screening and forking, or by electromagnetic apparatus if required. If small loose scrap is charged by fork rather than by shovel, small particles of iron or non-metallic particles should readily fall between the tines of a fork, whereas they would be retained by a shovel and go into the charge. Hard inclusions due to the brittle, complex, ferro-aluminium alloy which forms on the sides of iron pots may be largely prevented by thoroughly scraping the pots at frequent intervals and removing the accumulated scale. If 8- or 9-hour practice obtains, the pots should be scraped at least once a day, say at night, and preferably twice a day, at noon and night. If two shifts are working, more frequent scrapings will be advisable. All of the hard alloy should be removed so that the pot is clean, and if necessary the pot should be dumped. Moreover, during the time the pot is in melting operation, care should be taken that ladles or stirrers do not hit the sides of the pot; otherwise, some of the accumulated scale may be knocked off. In the same way, the scale may be dislodged at charging time.

By careful attention to the foregoing items, the occurrence of hard inclusions in aluminium-alloy castings that give rise to difficulty on machining and polishing can be largely eliminated.

CRACKS IN ALUMINIUM-ALLOY CASTINGS.

The common aluminium alloys are subject to cracking when poured into sand molds, and cracks constitute the most serious defect encountered in founding. Thus, on the basis that the average casting loss is 10 per cent for a variety of castings, the wasters, because of cracks, amount to about 2 per cent on the average of the castings poured, or 20 per cent of the total defectives. The question of cracks in aluminium-alloy castings has been discussed by the author [37] in detail in another place, and here it will be possible to deal with the subject only very briefly.

Roughly, a crack in a casting may be considered to be due to fracture of the alloy resulting from the stress set up by the contraction in volume on passing from the liquid state to the solid state. Both draws and cracks may be considered to be due to this cause. There are three kinds of cracks, viz., (1) external or surface cracks; (2) internal or deep-seated cracks; and (3) incipient or so-called strain cracks. An external crack is made evident as a large or small elongated opening in the surface of a casting. Such cracks may be very fine and tenuous or fairly wide, and, in aluminium-alloy castings, they may vary in length from about $\frac{1}{16}$ in. to 6 ins., or more. External cracks are found generally on inspection. Internal cracks are deep-seated fractures within the mass of the casting. An actual internal crack, although small, is a source of weakness; it will not be found on inspection of the rough casting, but it may be found after machining. Internal cracks, as well as blowholes, may be examined by the methods of radiography,[19] and X-ray examination will, in the future, be of great value in actual practice. Incipient cracks, more often known as strain cracks, can not be detected as such, in a casting, by any known method, since the term "strain crack" implies simply an internally (usually) stressed condition in the casting. These cracks are made evident, often, however, on machining, when the stresses and strains set up by machine work may be sufficient to cause a crack to appear. They may be induced by columnar crystallization in an alloy or by other conditions which lead to zones of weakness, or they may be due simply to the stresses which cause actual cracks.

Factors Affecting the Occurrence of Cracks.—In the case of sand castings, cracks have been attributed to a large number of items including the following: viz., improper methods of molding; wrong gating; too small, or an insufficient number of risers; chills left off, when required; chills put in the wrong place; too light or too heavy chills; excessively hot-short alloys; inferior aluminium pig, particularly pig high in both silicon and iron, or aluminium oxide; too high or too low pouring temperatures; hard ramming; dry-sand cores; core difficulties; wet sand; the presence of impurities in the alloy used; poor design of the casting, especially where there are heavy and light sections in contiguity or long thin webs between heavy bosses; and other causes. Broadly, cracks are directly due to the large contraction in volume of the aluminium-alloys on freezing, and the lack of its regulation or control. Examination of the general problem will show that the most important factors which have a bearing on the cracking of aluminium alloys on casting are the following: (1) contraction in volume; (2) composition of the alloy; (3) quality of the melting charge; (4) design of the casting; (5) method of molding; (6) hardness of ramming; (7) method of gating; (8) hardness and characteristics of cores; (9) chills; (10) risers; (11) melting temperatures; (12) furnace used for melting; (13) pouring temperatures; (14) inclusions in the alloys; (15) hot-shortness of the alloys; and (16) mechanical properties of the alloys at high temperatures. Of these sixteen subjects for consideration, the first, second, third, and eleventh to sixteenth involve ascertainable metallurgical facts, while the remainder are largely matters of opinion, judgment, and technical knowledge of men experienced in the production of castings. The effects of these various factors on the occurrence of cracks have been discussed by the author in another place,[37] and the confines of space permit only brief consideration of them here.

In general, the less the contraction in volume the less the tendency to cracking. The contraction is a function largely of the composition, and the fact that some aluminium alloys are more likely to crack than others is well known. Thus, the aluminium-zinc alloys crack more readily than the aluminium-copper alloys, while some aluminium-copper-zinc alloys (e.g., 85 : 3 : 12 aluminium-copper-zinc) crack less readily than the former. The silicon-bearing aluminium alloys have fairly low contraction, and these alloys are valued in casting practice because of this. The effect of the quality of the melting charge is important, and it is often alleged that cracks are traceable to poor quality of primary or secondary aluminium used for melting. So-called "brittle" aluminium pig, which may have cracks in the webs, has been found to yield cracked castings.

One of the more important factors affecting cracking is the design of the casting. As is well known, the presence of thick and thin sections in contiguity in a casting is a prolific cause for cracking, and Fig. 147 is a good example of a crack occurring at the juncture of such sections. The mechanism of cracking at the juncture of a thick and thin section may be discussed by reference to Fig. 148, which is a sketch showing a sharp change in section in a casting. It may be assumed for simplicity that aluminium is poured into the shape shown. On cooling from the liquid state, it may be considered that both the liquid and solidification shrinkage in part b has taken place while metal in the interior of part a is still liquid, i.e., shrinkage in part a occurs later than in part b. For simplicity, also, it may be assumed that the sections, if they be considered separately, contract toward their centers. The light section will freeze rapidly, but the heavy portion will freeze more slowly. Consequently, the complete casting will behave, during freezing, as though it con-

Fig. 147.—*Crack at the juncture of a thick and thin section in a crankcase; about ⅔ actual size.*

sisted of two different parts. The result is that the contraction of the light section is directed toward its center and that of the heavy section toward its center, and, during the freezing, the two sections will draw away from each other at the juncture x. The stresses set up by the solidification shrinkage are sufficient, in many cases, to develop fracture along the juncture of the two sections, but even if this is not the result, a serious plane of weakness is developed. The greater the contraction in volume of an alloy the greater the stress set up on solidification, and the greater the difference in size between the two sections the greater the cracking tendency. The design shown in Fig. 148 is mechanically weak. If, now, a rounded fillet be placed at the juncture of the thin and thick section, as shown in Fig. 149, then the bad effects of the design will be overcome in part, and the section will be less likely to crack. Of course, in practice, it is normally necessary to chill heavy sections to increase the rate of cooling.

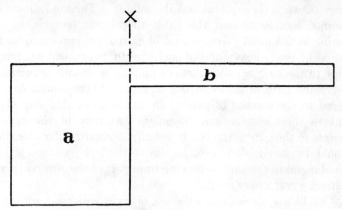

Fig. 148.—*Cracking at the juncture of a thick and thin section.*

The method and technique of molding is important in relation to cracking, and the general aspects of this may be indicated. Of course, the sand should be tempered properly, and it should not be too wet, as wet sand may cause cracks and warping. The bottom boards under the flask or mold box should be placed cor-

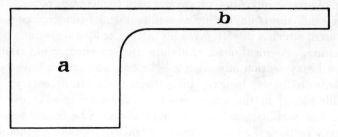

Fig. 149.—*Fillet at the juncture of a thick and thin section.*

rectly, and the box should be set evenly on the floor, since otherwise strains will be set up in the mold itself. The ramming should be even and of the correct hardness. If the sand is rammed too hard, the casting will be prevented from moving, while contracting on cooling, thereby setting up strains and, at times, causing actual fractures, i.e., cracks, in the casting. On the other hand, if the mold is rammed too lightly, there may be sudden collapsing of parts of the mold while the alloy is at a temperature immediately below that of final freezing, or in the range between the liquidus and the solidus. The sand should be rammed loosely around the gates to prevent hanging of the casting in the mold. In the case of gating, the sprue and leader should be made heavy and of such size that they will not freeze prior to the casting, and fillets of ample size should be provided where the gate attaches to the casting. Where cracks can be traced to the method of gating, an alteration of this may often lead to their elimination. Usually, that part of the casting nearest to the gate is the last to solidify, and particular attention should be given, therefore, to the method of gating. If still liquid metal in the gate is drawing from part of the already frozen casting, a crack may result.

Cracks are often due to the use of cores which are too hard, or which will not crush on contraction of the alloys, and according to Gillett,* this is one of the principal causes for cracks. A core should be so made that it will be sufficiently strong, yet sufficiently weak that it will crush readily under the stress set up by contraction of the alloy. Of course, cracks will always be less when green-sand cores are used, rather than dry-sand cores, although, if higher pouring temperature must accompany the use of green-sand cores, this would tend to increase cracks. The dry-sand core is unyielding as a rule, unless a binder, like resin, is used, which will soften readily at the temperature set up on pouring. As mentioned, chills are used to effect rapid cooling of a heavy section adjoining a light one, and cracks have been attributed to (1) lack of chills where chills are necessary; (2) chills placed in the wrong position in the mold or cores; (3) too heavy chills; and (4) too light chills. The failure to chill heavy sections is an usual cause for cracks. Speaking generally, it may be said that so far as risers are concerned in their relation to cracks, this defect has been attributed to (1) lack of a sufficient number of risers; (2) lack of sufficiently large risers, i.e., as to diameter and height; (3) too large risers; and (4) wrong position of the risers. Too many risers should not be used, because, under this condition, the casting may be anchored in the mold and not free to move, thereby causing cracking.

---

* Private communication, H. W. Gillett, Feb. 28, 1921.

Occasionally, cracks are attributed to so-called "burnt metal," i.e., overheating in the furnace, and this is likely because occluded dross particles, which would be more prevalent in overheated melts, would give rise to weak metal. Cracks have been attributed to too low and too high pouring temperatures, and that the pouring temperature is a factor is beyond dispute. An alloy poured at high temperature will be weaker than one poured at low temperature, and it will also have considerably greater total contraction in volume. The presence of foreign inclusions has been associated with cracking as has segregation of impurities. As will be explained later, cracks in both eutectiferous and solid-solution alloys have a strong tendency to be intergranular, and where foreign included matter occurs at grain boundaries (or elsewhere), it tends to weaken the alloy and aggravate any condition which may lead to cracks. Giolitti [29] appears to be convinced that the phenomena of local undercooling and segregation in steel ingots, and their effects on intercrystalline adhesion, are more important than internal stresses in originating flaws in steel, and he has found considerable segregation associated with cracks.

The hot-shortness of aluminium alloys is a direct measure of their tendency to crack, and other things being equal, alloys that are most hot-short crack most easily. The hot-shortness test is really a measure of the strength at elevated temperatures (in the solidification range). The tensile properties of alloys at elevated temperatures determine also the cracking tendency, and in general, the greater the strength and elongation the less the tendency to crack. The strength and elongation in the cold can not be taken as criteria of this tendency, nor can either of these properties be considered alone. Thus, certain aluminium-zinc alloys have enormous elongation at elevated temperatures, but low strength. These alloys crack more readily than some aluminium-copper alloys which have only moderate elongation at high temperatures, but still fair strength. On the whole, the elongation may be regarded as a poor measure of the cracking tendency, and, as determined by the ordinary tensile test at high temperatures, it will not reveal necessarily the presence of brittleness, even when present in very high degree. Extreme ductility under gradually and steadily applied tension may be accompanied by brittleness to sudden strains or shocks. Apparently, alloys that have high impact resistance at high temperatures will be less likely to crack than those with low resistance.

Metallography of Cracks.—Many samples of cracked aluminium-alloy castings have been examined by the author [37] with a view to determining the causes for cracking. Illustrative cases where cracks can be ascribed definitely to particular causes are not met with frequently in foundry practice, and this consequently beclouds study of the problem in its more practical

FIG. 150.—*Crack in 90 : 10 aluminium-copper alloy; vacuum-cleaner housing; unpolished;* × 6.

FIG. 151.—*Same as Fig. 150, but polished and etched NaOH; oblique illumination;* × 6.

aspects. The micro-metallography of cracks may be discussed conveniently by reference to a few typical cases.

Fig. 150 shows a crack in a vacuum-cleaner housing in 90 : 10 aluminium-copper alloy, and Fig. 151 shows the intergranular character of the same crack. Fig. 152 shows a crack (designated by X) in a section cut from an eight-cylinder motor crankcase, and Fig. 153 is a micrograph which shows the intergranular character of this crack. In studying the causes for cracking, a number of experiments were made by the author in the attempt to produce cracks at will in various aluminium alloys under definite conditions, but principally by pouring an annular-ring casting in a chill mold around a metal core. Although a crude method for determining the cracking tendency in different alloys, the expedient of pouring a casting in the form of an annular-ring has been of value in determining the die-casting qualities of aluminium alloys. It has been shown by preliminary experiments that the tendency to crack under this test

Fig. 152.—*Crack in a sample from a crankcase; actual size.*

varies with the alloy used and with the pouring temperature for a given annular-ring mold. Fig. 154 is a macrograph of a section in 85 : 3 : 12 aluminium-copper-zinc alloy poured in a small annular-ring mold at 870° C., showing the intergranular path of the crack.

In considering the microscopy of cracks, the question naturally arises: what path is taken by a crack, i.e., is it intergranular or intragranular? On theoretical grounds, it would be expected that a crack caused by stresses at high temperatures would follow

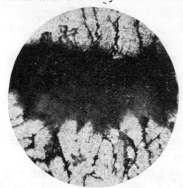

Fig. 153.—*Crack shown in Fig. 152; etched NaOH;* × 225.

the grain boundaries both in eutectiferous alloys and in solid-solution alloys. In the former, sufficient stress at high temperature will cause fracture through the eutectic areas. At a sufficiently high temperature, the eutectic will be liquid when the matrix of the alloy is still solid, since the eutectic is the alloy of lowest freezing point of any series, and under stress the fracture would occur through the weak liquid. Since eutectics are brittle, it would be expected that the fracture would take place through them rather than through the matrix, owing to contraction or to stresses at lower temperatures. In the case of

Fig. 154.—*Intergranular crack in chill-cast 85 : 3 : 12 aluminium-copper-zinc alloy; etched NaOH; oblique illumination;* × 3.

pure metals and solid-solution alloys, cracks might be expected to follow the grain boundaries under stresses at high temperatures.

As developed by Rosenhain from Beliby's original theory of the amorphization of metals on cold work, the grain boundaries of metals are presumed to be filled with an amorphous cement, this cement being the amorphous phase of the crystalline metal. This cement is claimed to be weaker than the crystalline metal above the equi-cohesive temperature but stronger below it. Whether the metal of the grain boundaries is actually amorphous or not has never been proved, but assuming for the purpose of the present case that it is amorphous, then at temperatures below the equi-cohesive temperature the fracture would be intragranular because the amorphous metal is stronger than the crystalline, but above the equi-cohesive temperature the amorphous phase is weaker than the crystalline phase, and consequently cracks owing to the solidification shrinkage will be intergranular. Where a crack is found assuming an indiscriminate path, it may be due to rupture at the equi-cohesive temperature, i.e., at that temperature where the resistance of the cement and of the grains is the same. It will have been noticed in certain of the accompanying macrographs that the grain size is large, and that the cracks passed through the boundaries of the large grains. In the eutectiferous 92 : 8 aluminium-copper alloy it is not to be supposed that the eutectic is segregated entirely at the boundaries of the large (macroscopic) grains, but each of the large grains will be found, on microscopic examination, to consist of a number of smaller grains, surrounded by the eutectic (cf. Chapter XVI). A crack may pass through a large grain and still be intergranular with respect to the small grains within the large grains (cf. Fig. 153).

Methods for the Prevention of Cracks.—In the discussion above, dealing with the effects of various factors on the occurrence of cracks, some methods for their prevention have been alluded to, and a consideration of the various factors will readily suggest methods by which cracks may be largely eliminated. Of course, it is not to be expected that cracks, as a defect, will entirely disappear by the observance of the precautions to be given below, but it is certain that they may be largely eliminated. In the following, the principal precautions to be observed may be pointed out briefly:

The contraction in volume is a fixed constant for a given alloy, and the use of alloys that have an excessively large contraction in volume should be avoided for the production of complicated castings or those of intricate design. Thus, certain aluminium-zinc alloys have a large contraction in volume, and such alloys will give much trouble if used for the production of large and complicated castings. The contraction in volume may be controlled within limits by the correct use of risers and chills, and a given alloy is not necessarily unsuitable for use in complicated castings simply because it has a high shrinkage. Both the contraction in volume and the mechanical properties at high temperatures are dependent upon the composition of the alloys, and alloys should be chosen, on the basis of adequate test data,[43] which properly fulfill the requirements. The quality of the melting charge should be controlled by chemical analysis; much of the difficulty experienced with cracking comes from inadequate control of the melting stock. Of course the castings should be designed correctly, both from the foundry and engineering points of view. So long as complicated and intricately designed castings must be made, it must be expected that cracks can not be wholly eliminated. From the foundry point of view, the simpler the design the less the difficulty experienced in production, and it would seem as though many castings made to-day are needlessly intricate. With regard to methods of molding and the use of chills and risers, these factors are variable, depending upon the casting made, and it is not possible to formulate any very definite rules. The expedient of experimental molding, as suggested by Traphagen,[21] is valuable in determining the most suitable method for molding a given casting, with particular reference to gating, risers, and chills. Too hard ramming should be avoided, and the required hardness of molds may be determined experimentally where experience is an insufficient guide. The gates should be of ample size and attached correctly to the casting; the use of horn gates where possible is recommended,[12] since they permit the alloy to enter the mold with the least disturbance. Much of the difficulty experienced with cracking is traceable to the cores and to faulty core-room practice, and too much emphasis can not be placed upon the necessity for using cores of the correct hardness. Chills are useful in preventing cracks, and many cracked castings could have been saved if they had been chilled properly. This is also true of the use of risers. While many simple castings do not require risers, it is a practical necessity to employ risers on most large and complicated castings so that certain parts of the castings may be fed with liquid metal.

The melting temperatures should be low, and overheating in the furnace should be avoided. Open-flame furnaces should be run with a non-oxidizing atmosphere in order to avoid the formation of much aluminium oxide, and consequently occluded dross in the alloy. Of course, the skimming should be adequately and carefully done, and the melting practice should be conducted as cleanly as possible. Foundry-floor sweepings should not be charged into the furnace unless sieved, and foreign materials should be kept out of the furnace charges. The pouring temperatures should be as low as is consistent with the metal filling the mold readily, and very high pouring temperatures should be avoided. This pre-supposes pyrometric control. The observance of these several precautions, together with proper supervision and the study of each cracked casting followed by remediable steps taken to prevent the occurrence in subsequent production, should go a long way in eliminating cracks as a serious defect in the founding of light aluminium-alloy castings.

INSPECTION AND SALVAGE SPECIFICATIONS FOR ALUMINIUM-ALLOY SAND CASTINGS.

Since casting losses are markedly affected by the rigidity of the specifications covering inspection, it has been deemed advisable to discuss the question of inspection in some detail, although at the necessity of sacrificing some important discussion on defects and their prevention. Various inspection instructions have been worked out in the larger foundries of the United States, and purchasers, such as motor companies, have drawn up specifications covering the allowable defects permitted in castings for use in motor cars. Of course, specifications suitable for covering the inspection of one kind of castings may be entirely unsuitable for another kind. Thus, inspection specifications for ordinary automotive castings would be entirely too lax for application to aircraft castings. The vendor and purchaser must decide to their mutual agreement as to what defects will be permitted in acceptable castings, and as to how serious a defect shall be so as to cause rejection of the casting. Necessarily, the variations permissible are wide, depending upon the type of casting and the use for which it is destined. Certain defects in one casting may cause rejection, but in another casting the same defects would be inconsequential or might admit of repair. One of the additional items affecting casting losses is that pertaining to salvage. Thus, if a defective casting can be salvaged by some method of repair, e.g., welding or soldering, then the casting will be acceptable, and no loss will accrue other than costs necessary to effect the salvage. The economic and metallurgical aspects of reclaiming defective castings by soldering and welding have been treated in detail by the present author and M. E. Boyd [47] (cf. also Chapter XVIII).

While it will be impossible here to detail inspection and salvage rules for all kinds of castings, the comprehensive set of

578 *Metallurgy of Aluminium and Aluminium Alloys*

inspection and salvage specifications,[27] drawn up by the Bureau of Aircraft Production, covering certain castings for the 12-cylinder Liberty aviation engine may be cited as a broad code of instructions. These specifications are more rigid than those usually enforced for automotive castings in general, but they serve to show the trend in requirements for the inspection and salvage of the highest class of aluminium-alloy sand castings poured. Specifications as to the inspection of aluminium-alloy sand castings for aviation engines both here and abroad during the World War were fairly rigid, and when consideration is given to the complexity of certain of these castings, especially crankcases, the claim of some foundrymen that they were too severe merits attention. In point of fact, certain aviation-engine castings were so poorly designed that it was practically impossible to put them into production without incurring heavy casting losses. The greater part of the specifications of the Bureau of Aircraft Production * for Liberty motor castings are reproduced below:

INSPECTION AND SALVAGE SPECIFICATIONS FOR LIBERTY AVIATION-ENGINE CASTINGS.

Vendor Inspection.—All castings made for the Liberty 12-cylinder aviation engine shall receive a very close inspection by the inspection department of the vendor foundry before being submitted to the government inspector (of the Bureau of Aircraft Production) for approval. This inspection shall consist of inspection of the castings on the foundry floor when removed from the molds, as well as the final inspection after the castings have been completely trimmed and are ready for shipment. After this inspection is made by the foundry inspectors, the castings shall be submitted to the inspectors of the Bureau of Aircraft Production for their approval. The vendor foundry will not be permitted to perform any repairing on castings before submitting them to the inspectors of the Bureau of Aircraft Production.

Inspection by the Bureau of Aircraft Production.—Metallurgical inspectors of the Bureau of Aircraft Production will inspect only those castings that have passed the vendor inspection. Castings that show small defects and that can be salvaged in accordance with the instructions given below will be salvaged under the direction of the inspectors of the Bureau of Aircraft Production. No deviation from these instructions will be allowed without written permission from the proper authority. The chief inspector for the Bureau of Aircraft Production at the plant of the vendor shall see that the material used is in accordance with the chemical and metallurgical specifications covering the castings and that the required number of physical tests are made.

Foundry Salvage.—All castings that show metallurgical defects shall be salvaged in accordance with the instructions given below. All castings for which these instructions do not give proper disposition, and that in the opinion of the inspectors of the Bureau of Aircraft Production can be salvaged, shall be held for the attention of the chief inspector on castings, who will make the proper disposition.

---
* These specifications are largely the work of Wm. R. Laird.

Welding.—The welding of all rough castings shall be carried out as follows: The entire casting shall be pre-heated to a temperature of approximately 260° C., or to a temperature at which the casting will not ring when tapped with a hammer. The defects to be welded must be thoroughly cleaned from any sand, oxide, or dirt, which would in any way interfere with the quality of the weld. The castings must be welded immediately after being removed from the pre-heating furnace. The welding stick used must be of approximately the same chemical composition as the casting itself. A finished weld must have the same contour as the original casting. The so-called "burning-in" of castings that show defects such as mis-runs or cracks, will not be allowed under any consideration. It will not be possible to weld any cracks on aluminium-alloy castings unless otherwise specified in detailed instructions listed below. No welding will be allowed on any ribs of upper and lower halves of crankcases. No welding will be allowed on propellor end of crankcase or oil pan beyond No. 7 rib.

Soldering.—No soldering of any kind will be permitted under any consideration on aluminium-alloy castings to be used on the Liberty motor.

Seepage Leaks.—Small seepage leaks which are detected by the open gasoline test may be treated by any of the following methods:

(a) Bakelite under pressure.
(b) Bakelite and aluminium powder.
(c) Boiled linseed oil under pressure—castings must be baked after this treatment.
(d) Accelerated aging process.
(e) Annealing of castings at a temperature of not over 316° C.

The use of aluminium powder and sulphur will not be permitted for the treatment of seepage leaks.

Blowholes.—Small blowholes or gas pockets which do not affect the strength of the casting and which do not leak when subjected to the open gasoline test shall not be a cause for rejection, and castings containing such flaws shall be accepted as they are.

Appearance.—No castings for use on the Liberty motor shall be treated or repaired in any way which will affect the appearance only.

INSPECTION AND SALVAGE OF CRANKCASES.

Inspection and salvage instructions for the crankcase (upper half) for the Liberty 12-cylinder motor are as follows with reference to Figs. 155, 156, and 157. The letters on the photographs refer to particular inspection points.

Referring to Fig. 155:
(A) Acceptance or rejection stamp shall be placed on this boss.
(B) Cracks caused by rough handling which run into crankcase flange screw hole are acceptable, provided crack does not extend beyond the hole.
(C) Castings showing porosity in this hole are acceptable, provided casting does not show oil leaks after assembly.
(D) Castings which show hole at this point drilled off center must be referred to the Production Engineering Department for disposition.
(E) Small cracks in ribs indicated, or small pieces broken from these ribs may be repaired by trimming rib with file until defect is completely removed and no sharp corners remain. Defects of this nature shall not penetrate any rib more than $\frac{3}{8}$ inch.
(F) Small cracks on this flange due to rough handling are acceptable, if $\frac{1}{8}$-in. hole is drilled at end of crack. No welding of cracks of this nature is acceptable.

As applied to Fig. 156:
(A) Great care shall be taken to see that no castings are accepted showing shrinkage cracks under thrust-bearing retainer and propellor-hub bearing seat.

(B) Castings which show bearing-felt-retainer inner wall not less than $\frac{1}{16}$ in. thick are acceptable. Castings showing corners undercut, so that thickness of wall between corners and bolt holes at this point is not less than $\frac{1}{16}$ in. thick, are acceptable.

(C) Castings which show holes drilled too near edge of flange are acceptable, providing no oil leaks occur after assembly.

FIG. 155.—*Top view, Liberty 12-cylinder aviation-engine crankcase, propellor end left.*

(D) Castings showing leaks in No. 7 bearing by the open gasoline test indicate presence of blowholes under the bearing and they shall be rejected.

(E) Cracks which penetrate the flange at this point and stop are acceptable as they are. No welding of this type of defect is permissible.

FIG. 156.—*Bottom view, Liberty 12-cylinder aviation-engine crankcase, propellor end left.*

(F) Cracks which penetrate the flange at this point and stop are acceptable as they are. In case the crack does not run into a hole, then $\frac{1}{8}$-in. hole should be drilled at the end of the crack. No welding is permissible on this type of defect.

(G) Cracks which penetrate to a point not beyond the center line of holes are acceptable when a $\frac{1}{8}$-in. hole is drilled at the end of the crack. No welding of defects of this kind is permissible.

Casting Losses and Defects in Castings

(H) Castings showing porosity on surface are acceptable, providing they do not show oil leaks after assembly, or after having completed acceptance run. No shellac or other foreign material shall be used for stopping these leaks.

(I) Castings showing holes or mis-runs in webs indicated shall be rejected.

(J) Cracks in ribs indicated are not acceptable.

(K) Cracks at points indicated are not acceptable.

(L) Cracks in bosses are acceptable, provided they are not over $\frac{3}{8}$ in. long. In case two cracks occur on the same boss and are not more than $\frac{1}{8}$ in. apart, section between cracks shall be removed. A maximum of two cracks will be allowed to a single boss, and there shall not be more than one defective boss to a cylinder.

(M) Depressions in surface due to core fins which are not over $\frac{1}{16}$ in. deep are acceptable. In every case, such depressions must be thoroughly cleaned.

(N) Sand holes in this face are acceptable, provided they do not leak on the open gasoline test. In case small sand holes are found that do leak,

FIG. 157.—*Top view, Liberty 12-cylinder aviation-engine crankcase, propellor end right.*

these holes may be repaired by the use of a $\frac{1}{8}$-in. aluminium pipe plug. The plug must have at least three effective threads, and it must be screwed in tightly and cut off flush with each side of the casting. No peaning will be allowed. After being repaired the casting shall undergo an open gasoline test for leaks.

(O) Sand holes on this face are acceptable, provided they are thoroughly cleaned and do not show oil leaks.

(P) Cracks which penetrate to this depth are acceptable, provided they run into bolt hole. Cracks which do not run into bolt hole or do not extend beyond the centre line of the hole are acceptable after a $\frac{1}{4}$-in. hole is drilled through the flange at the end of the crack.

The following apply to Fig. 157:

(A) Castings showing porosity on this face are acceptable, provided no oil leaks occur after assembly.

(B) Castings which show porosity on the surface indicated are acceptable, provided no oil leaks occur after assembly.

(C) Castings showing small cracks at this point due to use of oversized stud are acceptable, provided they are less than $\frac{3}{8}$ in. long and not more than one crack occurs on each cylinder pad.

582 *Metallurgy of Aluminium and Aluminium Alloys*

INSPECTION AND SALVAGE OF OIL PANS.

Inspection and salvage instructions for the oil pan (lower half) for the Liberty 12-cylinder motor are as follows, with reference to Figs. 158, 159, and 160. As before, the letters on the photographs refer to particular inspection points.

Referring to Fig. 158:
(A) Holes in ribs which, after being thoroughly cleaned, are not over $\frac{3}{8}$-in. diameter or equivalent area, and which are located at least $\frac{1}{2}$ in. from fillet, are acceptable.

(B) Holes which are located at least 1 in. from ribs are acceptable when such holes can be repaired with a $\frac{1}{8}$-in. aluminium pipe plug. The plug must have at least three effective threads, and it shall be screwed in tightly and cut off flush with each side of the casting. No peaning will be permissible. After being repaired, the castings shall undergo an open gasoline test for leaks.

(C) No shrinkage cracks at this point are permissible. Small cracks, due to the chipping chisel striking the wall on chipping, which occur on the

Fig. 158.—*Top view, Liberty 12-cylinder aviation-engine oil pan, propellor end left.*

outside of the casting at this point, are acceptable as they are, provided the casting will stand the open gasoline test for leaks. No peaning of cracks is permissible.

(D) Castings which show not more than two small cracks on any single boss, due to tightening of the bearing bolt, are acceptable. Castings which show cracks at this point, in the rough condition, are defective, and they must be scrapped.

(E) Castings on which corners of bearings show sand holes or small nicks are acceptable provided these are cleaned properly and do not extend more than $\frac{3}{8}$ in. from the end of the bearing and not more than $\frac{3}{8}$ in. along the side of the bearing. Castings showing porous bearings are acceptable, provided sand holes are not more than $\frac{1}{8}$-in. diameter. The holes shall not make up more than 10 per cent of the surface area of the bearing, and they shall not be more than $\frac{1}{8}$ in. deep. In case only one sand hole is found on a single bearing, and the diameter of the hole is not more than $\frac{1}{4}$ in. the casting shall be accepted.

(F) Small cracks in the flange caused by rough handling are acceptable, provided the crack is at least $\frac{3}{8}$ in. from the base of the radius and does not run beyond the center line of the bolt holes, if a $\frac{1}{8}$-in. hole is drilled at the end of the crack. Cracks which extend beyond this point can not be repaired on finished castings. However, if cracks are present on rough or semi-finished castings and do not extend beyond this point or down the side of the casting to a depth not to exceed the width of the flange, such cracks may be repaired by drilling a ⅛-in. hole at the end of the crack and then welding.

(G) Castings which show porosity or sand holes on the surfaces indicated are acceptable, if thoroughly cleaned to a bright surface, provided no leaks will occur at these points after assembly.

FIG. 159.—*Top view, Liberty 12-cylinder aviation-engine oil pan, propellor end right.*

Then referring to Fig. 159:

(A) Castings which show bearing-felt retainer inner wall not less than $\frac{1}{16}$ in. thick are acceptable. Castings showing corners undercut so that the thickness of the wall between the corners and the bolt holes at this point is not less than $\frac{1}{16}$ in. thick are acceptable.

(B) Castings showing shrinkage cracks under thrust-bearing retainer shall be scrapped. Extreme care shall be taken to see that no shrinkage cracks occur at this point.

(C) Castings showing general porosity over finished surface are acceptable, provided no oil leaks occur after assembly, or after completing acceptance test. No shellac or other foreign material shall be used to stop such possible leaks.

(D) Sand holes, mis-runs, and small cold shuts are acceptable at this point, provided such defects are not more than ½-in. diameter.

(E) Small cracks which occur on bosses at this point are acceptable, provided not more than two cracks occur on a single boss.

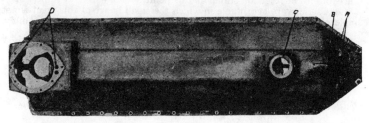

FIG. 160.—*Bottom view, Liberty 12-cylinder aviation-engine oil pan, propellor end right.*

The following apply to Fig. 160:

(A) Porosity and sand holes in bolt holes, or spot-faced surfaces, are acceptable when thoroughly cleaned, provided the casting does not leak at this point and that the depth and extent of the defect is apparent.

584 Metallurgy of Aluminium and Aluminium Alloys

(B) Defects which occur at this point shall not be repaired by welding.

(C) Castings which show porosity on this machined surface are acceptable, provided no leaks occur after assembly.

(D) Rough castings which are light of stock on this flange shall be repaired by welding stock on the flange. Castings which show porosity on the finished surface are acceptable, provided no leaks occur after assembly.

(E) Minimum wall thickness of castings, variable because of core shifts, shall be $\frac{1}{8}$ inch.

INSPECTION AND SALVAGE OF CAMSHAFT HOUSINGS.

Inspection and salvage instructions for camshaft housings for the Liberty 12-cylinder aviation engine are as follows with reference to Figs. 161, 162, 163, and 164. The letters on the photographs refer to particular points of inspection.

Fig. 161.—*Bottom view, Liberty 12-cylinder aviation-engine camshaft housing.*

The following refers to Fig. 161:

(A) Shrinkage cracks at or adjacent to fillet at this point are acceptable, provided such cracks can be completely filed out or scraped out, and provided, further, the remaining section is not diminished more than $\frac{1}{32}$ in. A series of sand holes in this part of the casting is cause for rejection.

(B) All porous spots on this surface must have all foreign matter removed. Castings with porosity making up more than 10 per cent of the bearing surface shall be rejected. Porous spots shall not be more than $\frac{1}{8}$ in. wide.

(C) Castings showing porous finished surfaces are acceptable, provided the porosity is not sufficient to cause leaks after assembly.

(D) Small shrinkage cracks or stress cracks are acceptable at this point, provided they can be removed by filing or scraping, and the section is not diminished thereby more than $\frac{1}{64}$ inch.

(E) No cracks of any kind are acceptable at this point.

(F) All castings which are salvaged shall be properly marked with either an acceptance or rejection stamp at this point.

(G) Castings are acceptable in which the small flange at this point is not completely filled out, provided the castings will show no leaks after assembly.

Fig. 162.—*Top view, Liberty 12-cylinder aviation-engine camshaft housing.*

Considering Fig. 162:

(A) Castings which show thin flange at this point are acceptable, if the flange is not less than $\frac{3}{16}$ in. thick.

(B) Castings which show porosity on this face are acceptable, provided no leaks occur after assembly.

(C) Small shrinkage cracks at this point are acceptable, if such cracks can be completely removed by scraping or filing and the section is thereby diminished not more than $\frac{1}{64}$ inch.

(D) This point must not be peaned, and it must not leak when subjected to the open gasoline test.

(E) Great care must be taken to avoid cutting main body with trimming chisel.

(F) Small depressions on core fins at this point are acceptable, provided the depression is not more than $\frac{1}{32}$ inch.

(G) Castings showing thin wall at this point are acceptable, providing the wall at the thinnest portion is at least $\frac{3}{32}$ in. thick.

(H) Small cracks due to straightening are acceptable if such cracks can be completely removed by filing or scraping without reducing the thickness of the wall at this point more than $\frac{1}{32}$ in. Sand holes occurring on the body of the housing must not be more than $\frac{3}{64}$ in. deep.

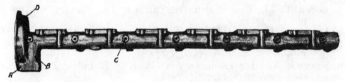

FIG. 163.—*Side view, Liberty 12-cylinder aviation-engine camshaft housing.*

Referring to Fig. 163:

(A) Small shrinkage cracks at this point on rough castings may be welded after being thoroughly cleaned.

(B) Castings which show small leaks due to sand holes at this point may be plugged with a $\frac{1}{8}$-in. aluminium pipe plug. The plug must have at least three effective threads, and it shall be screwed in tightly and then cut off flush with each side of the casting. No peaning will be allowed. After being repaired, the casting shall undergo an open gasoline test for leaks.

(C) Holes which are eccentric at this point are acceptable, provided the thickness of the wall at the thinnest portion is not less than $\frac{1}{16}$ inch.

(D) Small cracks due to trimming inside of bell are acceptable, provided the cracks do not leak on the open gasoline test.

As to Fig. 164:

(A) Castings showing porosity on this surface are acceptable, provided no oil leaks occur after assembly.

(B) Castings showing thin walls are acceptable, provided the thinnest portion is not less than $\frac{1}{16}$ in. thick.

Specifications for the inspection and salvage of pistons for the Liberty aviation engine are also included in the instructions of the Bureau of Aircraft Production.

FIG. 164.—*End view, Liberty 12-cylinder aviation-engine camshaft housing.*

SELECTED BIBLIOGRAPHY.

The following references include a number not cited in the chapter but which have a bearing on the general subject of defects in aluminium-alloy castings. Reference may also be made to the bibliography appended to Chapter XI.

1. Dumas, M., Sur les gaz reténus par occlusion dans l'aluminium et le magnésium, *Compt. Rend.*, vol. 90, 1880, pp. 1027–1029.
2. Fuller, B. D., The gating of castings, *The Foundry*, vol. 22, 1903, pp. 13–17.
3. Bedell, F. M., Cost of defective castings, *The Foundry*, vol. 23, 1903, pp. 130–131.
4. Leary, W., Crush of the mold, *The Foundry*, vol. 24, 1904, pp. 165–166.
5. Desch, C. H., Some common defects occurring in alloys, *Jour. Inst. of Metals*, vol. 4, 1910, pp. 235–246.
6. Sieverts, A., Über Lösungen von Gasen in Metallen, *Zeit. für Elektrochem.*, vol. 16, 1910, pp. 707–712.
7. Sieverts, A., and Krumbhaar, W., Über die Löslichkeit von Gasen in Metallen und Legierungen, *Ber. Deut. Chem. Gesell.*, vol. 43, 1910, pp. 893–900,
8. Guichard, M., and Jourdain, P.-R., Sur le gaz de l'aluminium, *Compt. Rend.*, vol. 155, 1912, pp. 160–163.
9. Perkins, J. M., Defective castings—how to handle them, *Trans. Am. Foundrymen's Assoc.*, vol. 20, 1912, pp. 355–361; and *The Foundry*, vol. 38, 1911, pp. 276–277.
10. Anon., Use of green sand cores in aluminium work, *The Foundry*, vol. 40, 1912, pp. 141–146.
11. Gillett, H. W., Influence of pouring temperature on aluminium alloys, Eighth Intern. Cong. Appl. Chem., vol. 2, 1912, pp. 105–112.
12. Collins, J. W., Suggestions for making aluminium castings, *The Foundry*, vol. 42, 1914, pp. 67–68; abst. of paper before the Detroit Foundrymen's Assoc.
13. Desch, C. H., The solidification of metals from the liquid state, *Jour. Inst. of Metals*, vol. 11, 1914, pp. 57–118; and *idem*, vol. 22, 1919, pp. 241–263.
14. Davey, W. P., Application of the Coolidge tube to metallurgical research, *Gen. Elec. Rev.*, vol. 18, 1915, pp. 134–136.
15. Tonamy, C. H., Detection of internal blowholes in metal castings by means of X-rays, *Jour. Inst. of Metals*, vol. 14, 1915, pp. 200–203.
16. Davey, W. P., Radiography of metals, Trans. Amer. Inst. of Min. Engrs., vol. 53, 1916, pp. 150–160.
17. Carpenter, H. C. H., and Elam, C. F., An investigation on unsound castings of admiralty bronze (88 : 10 : 2); its cause and the remedy, *Jour. Inst. of Metals*, vol. 19, 1918, pp. 155–175.
18. Comstock, G. F., Notes on non-metallic inclusions in bronze and brass, Trans. Amer. Inst of Metals, vol. 12, 1918, pp. 5–10.
19. Symposium, Examination of materials by X-rays, Trans. Faraday Soc., vol. 15, 1920, pp. 1–82; papers by R. Hadfield, W. H. Bragg, E. Schneider, A. W. Porter, *et al.*
20. Anderson, R. J., Blowholes, porosity, and unsoundness in aluminium-alloy castings, U. S. Bureau of Mines Tech. Paper 241, November, 1919, 34 pp.
21. Traphagen, H., The value of a scrap pile, paper before the Amer. Foundrymen's Assoc., Philadelphia meeting, September, 1919.
22. Anderson, R. J., Metallography of aluminium ingot, *Chem. and Met. Eng.*, vol. 21, 1919, pp. 229–234.

23. Lea, F. C., The founding of aluminium, *The Metal Ind.* (London), vol. 15, 1919, pp. 509–511; abst. of paper before the Royal Aeronautical Soc., April, 1919.
24. Anon., Aluminium-foundry molding losses analyzed, *The Foundry*, vol. 47, 1919, p. 416.
25. Anderson, R. J., Unsoundness in aluminium castings, *The Foundry*, vol. 47, 1919, pp. 579–584.
26. Anderson, R. J., and Capps, J. H., Investigation of hard spots in aluminium, *The Foundry*, vol. 48, 1920, pp. 337–342.
27. Anderson, R. J., Analysis of losses in aluminium shops, *The Foundry*, vol. 48, 1920, pp. 989–992; and vol. 49, 1921, pp. 16–21; 54–57; 104–111; 143–147; 188–191; and 235–239.
28. Anderson, R. J., Casting losses in aluminium-foundry practice, U. S. Bureau of Mines Reports of Investigations, April, 1920, Section VIII.
29. Giolitti, F., Cracks in ingots, *Chem. and Met. Eng.*, vol. 23, 1920, pp. 149–153.
30. Gibson, W. A., Position of tensile tests in the foundry, *The Iron Age*, vol. 105, 1920, pp. 725–728.
31. Bezzenberger, F. K., and Wilkins, R. A., Apparatus for the determination of the porosity of metals, *Chem. and Met. Eng.*, vol. 22, 1920, pp. 1031–1032.
32. Anderson, R. J., Castings of light aluminium alloys, *The Iron Age*, vol. 107, 1921, pp. 433–436.
33. Rawdon, H. S., Macroscopic examination of metals, *Chem. and Met. Eng.*, vol. 24, 1921, pp. 385–387.
34. Turner, T., The casting of metals, *Jour. Inst. of Metals*, vol. 26, 1921, pp. 5–43.
35. Anderson, R. J., Casting losses in aluminium-foundry practice, Trans. Amer. Foundrymen's Assoc., vol. 29, 1921, pp. 459–487.
36. Anderson, R. J., Iron-pot melting practice for aluminium alloys, *The Metal Ind.*, vol. 19, 1921, pp. 189–190; 246–247; 285–287; 360–362; 397–399; and *idem*, vol. 20, 1922, pp. 60–61; 309–311.
37. Anderson, R. J., Cracks in aluminium-alloy castings. Trans. Amer. Inst. of Min. and Met. Engrs., vol. 68, 1922, pp. 833–860.
38. Anderson, R. J., Inclusions in aluminium-alloy sand castings, U. S. Bureau of Mines Tech. Paper 290, June, 1922, 25 pp.
39. Rosenhain, W., and Grogan, J. D., The effects of overheating and repeated melting on aluminium, *Jour. Inst. of Metals*, vol. 28, 1922, pp. 197–213.
40. Czochralski, J., Die Löslichkeit von Gasen im Aluminium, *Zeit. für Metallkunde*, vol. 14, 1922, pp. 737–741.
41. Czochralski, J., Fremdstoffeinschlüsse im Aluminium, *Zeit. für Metallkunde*, vol. 15, 1923, pp. 273–283.
42. Botta, R., Peening or impregnating porous aluminium castings, Bureau of Aeronautics, Navy Dept., Tech. Note No. 109, Dec. 10. 1923.
43. Anderson, R. J., Linear contraction and shrinkage of a series of light aluminium alloys, Trans. Amer. Foundrymen's Assoc., vol. 31, 1924, pp. 392–466.
44. Lyon, A. J., Hard spots in aluminium alloys, *The Foundry*, vol. 52, 1924, pp. 396–397.
45. Lobley, A. G., Non-metallic inclusions in aluminium, *The Metal Ind.* (London), vol. 24, 1924, pp. 474–475; abst. of paper before The Faraday Soc.
46. Anderson, R. J., and Boyd, M. E., The production of aluminium-alloy pistons in permanent molds, paper before the Amer. Foundrymen's Assoc., Milwaukee meeting, Oct., 1924.
47. Anderson, R. J., and Boyd, M. E., Salvage and reclamation of aluminium-alloy castings by soldering and welding, paper before the Amer. Foundrymen's Assoc., Milwaukee meeting, Oct., 1924.

CHAPTER XIII.

PRODUCTION OF DIE CASTINGS AND PERMANENT-MOLD CASTINGS.

This chapter is divided into two parts, the first dealing with the production of die castings in aluminium alloys, and the second with permanent-mold castings. It will not be possible here to deal with the intricate details of the manufacturing processes as applied to the production of die castings and permanent-mold castings, but rather, it is the author's object to give only a broad survey of the production of these two types of castings. The die-casting process has been expanding rapidly in recent years, and a great variety of parts are now made by die casting. The permanent-mold process has been so far largely limited to the production of pistons for internal-combustion engines, but an increasing variety of castings is now being made by casting in permanent molds. The principles involved in the production of castings in metal molds as contrasted with sand molds have been known and practised for many years, but the creation of die casting as a separate industry dates back only to about 1900. The permanent-mold casting of metals and alloys, however, antedates sand casting by many hundreds of years, and stone and bronze molds were employed for making simple castings by prehistoric man. The history of permanent-mold casting has been discussed by Johnson,[32] and by the author and M. E. Boyd.[42] Of course, the extensive use of permanent molds for the production of castings is a comparatively recent achievement, and the use of permanent-mold castings has not grown so rapidly as that of die castings.

The use of metallic molds for the production of castings finds justification in the fact that practically all castings produced in sand molds require more or less machining, and that, by the use of a metallic mold of some kind, castings which are more accurate and which require less machining can be made. The question of machining costs is especially important in the production of modern metal manufactures. Practically any mechanical part, unless too large, can be made by die casting or permanent-mold casting, and castings made by these methods may be practically finished parts when they come from the mold, except for the removal of fins and some trimming and shaving that may be necessary on some surface. The importance of die casting in modern manufacturing is generally recognized in the metal industries, and the economies that can be effected by the use of die castings in place of sand castings or worked parts are well known. Five different processes are in use for the production of castings which involve the use of metallic molds, viz., (1) die casting; (2) permanent-mold casting; (3) the Cothias system; (4) centrifugal casting; and (5) slush casting. The first two will be discussed here, but the Cothias process need be given only brief consideration because it is so far not important for aluminium-alloy work, while the production of slush castings is confined solely to low-melting point alloys, and need not be taken up at all. Aluminium alloys are not cast centrifugally in practice.

PRODUCTION OF DIE CASTINGS.

The manufacture of aluminium-alloy die castings became important in 1914, all die castings made before then being of tin-rich, zinc-rich, or lead-rich alloys. Modern die-casting practice had its origin apparently in type founding, as exemplified by the machine invented by Bruce in 1838. In 1855, Mergenthaler brought out the linotype machine, which employs the recognized principles of die casting. At the outset, it is important to state clearly what is meant by a die casting, because there is great confusion in the use of terms. Die castings are, first of all, not to be confused with die pressings which are made by the hot-pressing of various metals and alloys. Die pressings are often referred to as die-pressed castings and die castings. Permanent-mold castings are often called "die castings." *A die casting is defined as a finished casting made by forcing a liquid metal or alloy by pressure into a metallic mold or die.* It is supposed that little or no machining, other than drilling for screws, bolts, and the like, will be required to put the casting into condition for use.

A permanent-mold casting is a product obtained by pouring a liquid metal or alloy into a metallic mold, the metal going into the mold under the force of gravity. The distinction between die castings and permanent-mold castings is thus made plain by definition. In the Cothias process, liquid alloy is forced into a mold by mechanical pressure, the plunger conforming in shape to the inside of the casting and acting as the core. Castings made by this process may be regarded as die castings. Slush castings are made simply by pouring a low-melting point alloy into a mold, and then immediately pouring out the unfrozen interior, leaving a hollow casting formed by the contour of the mold. A die casting is a *finished casting* in the general meaning of the term, in contradistinction to a rough sand casting for which a greater or lesser amount of machining is invariably presupposed.

Die casting is essentially a quantity production process, and but few parts can be considered practical die-casting jobs in less than 1,000 lots. This is necessarily so because of the heavy expense involved in designing and making the requisite dies; in small lots, this expense will be proportionally reflected in the cost of the castings. Where lots of 10,000 or more castings are run, the die cost will be readily absorbed and actually make up but a small part of the cost of the casting. The cost of any die casting must be less than the combined cost of producing a sand casting, plus the cost of machining to a finished product equivalent to a die casting. The die-casting process, as applied to aluminium alloys, is limited now to castings not longer than 24 ins. and weighing not more than 10 lb., but the process is being developed continually to include larger and heavier castings. The process is especially applicable to the production of small interchangeable parts which do not need to be especially strong, but are required to be well finished and accurate as to dimensions. Thus, the principal advantages of die casting over sand casting are as follows: (1) die castings are accurate and uniform; (2) machining costs are either eliminated entirely or greatly reduced; (3) the process is continuous, and the output is usually more rapid than could be obtained for the same part by sand casting; and (4) parts that could not possibly be sand cast can be successfully die cast. The main function of die casting is to save machining costs, and where this is not important, a die casting is not ordinarily made. Frequently, however, an irregularly shaped part, which has been sand cast in several pieces so as to permit machining, may be re-designed and cast as a single piece by the die-casting process. Scherer [7] has well illustrated the advantages of die casting from the manufacturing point of view in reducing the number of operations as compared with sand casting; the production of a small motor bracket with a bronze insert was considered. If sand cast, the following operations would be necessary: molding; grinding, and cleaning; turning, facing, and boring; drilling bolt and oil holes; drilling and reaming for brush holder; and pressing in bearing. When die cast, the operations are: molding; cleaning and removing fins; and reaming holes where necessary. On the other hand, die castings have some very definite disadvantages which should not be overlooked in considering the application of a die-cast part for any given purpose. Die castings in aluminium alloys made on compressed-air machines are exceptionally unsound and full of blowholes, and in fact, no actually sound aluminium-alloy die casting has been made. The outer shell of the casting has a good, sound surface appearance, but if that shell is removed by machining, the unsound and blowholey interior will be revealed. Moreover, the castings are not strong, because of this condition, and die castings should certainly not be used where soundness and strength are essential. Permanent-mold castings in aluminium alloys have better properties than either die castings or sand castings, both from the point of view of strength and soundness, and often where die casting is contemplated it will be found advisable to make parts by the permanent-mold process.

Aluminium-alloy die castings can be made to specifications of ± 0.005 in., and under some conditions,[3] if necessary, to ± 0.0005 in. For the 90 : 10 aluminium-copper alloy, the following standard limits are set by the Doehler Die-Casting Co.: maximum weight for castings, 8 lb.; limits in wall thickness, $\frac{1}{16}$ in. for small castings and $\frac{1}{8}$ in. in large castings; variations from drawing dimension, ± 0.0025 in. of diameter or length; maximum number of threads, external threads, 24 per in., limits on threads ± 0.004 in. (internal threads not cast); holes, 0.093 in. minimum, not deeper than $\frac{1}{8}$ in. (smaller holes can be spotted to facilitate drilling); draft, cores, 0.015 in. per in. of length and diameter (side walls, 0.005 in.), and cores less than $\frac{1}{4}$ in. diameter to have 0.005 in. per in. of length and diameter.

Mode of Manufacture of Die Castings.—The process of die casting consists essentially in melting the alloy in a suitable container and then forcing it by suitable means under pressure into a metallic mold or die. The resultant casting is smooth as to surface, requires little or no machining, and is ready for buffing or plating after the removal of fins and dressing. As outlined by Pack,[11] the following three items are essential to successful die casting: (1) a metallic mold or die capable of withstanding the action of the liquid alloy, as well as rapid temperature changes; (2) a machine or appliance for delivering the liquid alloy into the die or mold under pressure sufficient to ensure a casting of perfect contour; and (3) an alloy suitable for the process. The essential steps in the production of finished die castings are these: (1) design of the casting and die, which must be properly co-ordinated; (2) set-up of the die-casting machine; (3) delivery of the alloy to the melting pot; (4) casting of the alloy in the die-casting machine, as described below; (5) inspection of the castings for defects; (6) cleaning, where necessary; (7) plating, if required; and (8) polishing, if necessary.

For convenience, the process of die casting can be described by outlining the operation of a typical machine. Fig. 165 is a transverse section of the Hakanson die-casting machine, which is typical of the successful machines applied to aluminium alloys, and Fig. 166 is a similar section with the movable parts in different operative positions. In Fig. 165, *1* is a vertical die plate, pivotally supported upon a stand *28* by passing through a shaft *29* through the holes provided for that purpose. Near the corners of the die plate, four tie-rods * *16*, are secured, connected at their other extremities to the casting of the pneumatic motor *21*. The latter is supported upon its lower extension *37* by the pivot, *43*, which is in turn supported in the base *30*. The nut above the base forms in combination with a thread on the pivot, a means for adjusting the height of the motor end of the machine. A movable die plate *18* is slidably mounted on the tie-rods, *16* and it is connected to the arms *20*, or motor *21* by the links *19*. The intake at *36* is connected to an air tank or compressor. If the lever *31* is drawn a quarter turn toward the operator, the slots in the valve will reverse connections

---

* The plural is used in several cases in this description, as only the cross-section of the machine is shown.

with the motor, and the air from pipe *36* will pass into the upper channels of the motor casting *21*, thereby rotating an internal sector and causing exhaustion of the air through the vent *35*. Since the arms *20* are keyed to the shaft of the sector, they will now have moved to the position shown in Fig. 166, as contrasted with that shown in Fig. 165, the links *19* having drawn the movable die plate toward the motor and opened the die *17*. If it is desired to close the die, the valve lever *31* is pushed back to its former position. The motor *21* is a semirotary oscillating pneumatic motor, and it is milder in its action at the end of the stroke when closing the die than a direct-acting hydraulic piston; the latter always imparts the same speed to a movable die plate throughout the stroke, and tends to terminate the stroke in a violent blow. A furnace *25* is rolled up to the stand *28* as far as possible. The melting pot *24* is supported in the furnace by small studs or cleats. The method of heating is preferably by gas or oil.

A block *8* is secured to the die plate *1* and is provided, upon its upper portion, with a rearward extension over the shoulder. The top of the die plate has a set-screw *45* which forms contact with the bottom of the shoulder, and by means of this screw, block *8* and all parts may be adjusted in height. A hollow shaft, bearing an air valve *3*, passes through the upper portion of the block *8* and supports the upper link *1'* and *10*. From the lower part of the same block, a pair of extensions project forward and support the shaft rotated by lever *22*. This latter shaft provides a pivot for the links *6*, and has a short arm *o* rigidly secured to it. The outer ends of the links *1'*, *10*, *6*, bear shafts which pass through the upper portion of web *2*; the ladle *12* is attached to this web, the upper hollow shaft passing through one hub and the longer shaft through another hub. Further connection to the web *2* above the ladle is made by adjustable links *5* whose shaft *4* passes through a hub; the inner ends of these links are connected to the arm *o* of the lever shaft. If the lever *22* is now drawn slowly forward and down until it reaches a position shown by Fig. 166, arm *o* will draw down the ladle from die plate *1* by means of the links *5* to the position shown in this figure. This resultant position of the ladle is brought about by the slightly different length of the upper and lower pairs of links, causing the nose of the ladle to dip sufficiently under the surface of the liquid alloy so as to avoid being charged with any dross floating on top. Upon being raised from the bath of liquid alloy, whatever dross may have entered, will pour out before the nose of the ladle meets the die. The construction of the ladle provides for an integral web between the ends, thus rendering it unaffected by heat, as the bearings in the web extension and the parts composing the ladle mechanism are out of range of the heat from the pot. The blocks in the links *5* may be adjusted by set-screws, thus adjusting the final position of the ladle.

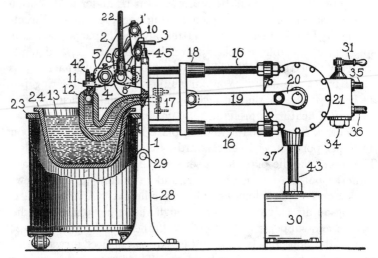

FIG. 165.—*Side view of the Hakanson die-casting machine; ladle in contact with die opening (Hakanson).*

In order to force the liquid alloy into the die, connection is made by piping, not shown in the figures, to the air valve *3*. When the lever end of this valve occupies the rearward position shown in Fig. 165, the air passes through the shaft upon which the valve is secured, through link *10* and into the link *1'*. Thence, the air rises upward in the same link and out into a tee, down through a pipe and into the large end of the ladle. When the valve is closed as in Fig. 166, and the ladle lowered into the melting pot, the stopper *42* will be released from the vent in the ladle, removing the pressure from the contained alloy. The operation of this stopper is controlled by a cam on the shaft of lever *22*, the cam being adapted so as to engage the inner ends of the stopper arm *11* when the lever is raised to a vertical position. In operating the machine, the starting position may be taken as illustrated in Fig. 165. First, the die is closed by pushing the lever *31* to the right. Then, the lever *22* is drawn around clockwise two-thirds of a revolution into a vertical position until the nose of the ladle meets the receiving end of the die; then, the air is turned on at valve *3*. The liquid alloy is immediately forced into the die and freezes almost at once. Valve *3* is turned off, the ladle is then lowered into the melting pot by means of the

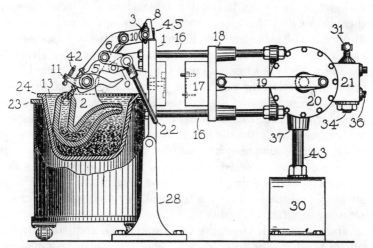

FIG. 166.—*Side view of the Hakanson die-casting machine; ladle dipping into bath and movable parts in different operative positions from Fig. 165 (Hakanson).*

lever *22*, and the die is opened by turning on the valve *34*. When the casting has been removed, the above operations are repeated in the same order indefinitely. The melting pot and furnace are separate from the die-casting machine proper, thereby avoiding conduction of heat.

In operating die-casting machines, the question of die temperatures is exceedingly important, and in practice, dies are normally run at 350 to 500° C., depending upon the type of casting and the rapidity of operation. When a machine is started from the cold, many defective castings may be made until the die is hot, and the maintenance of a fairly uniform temperature in the die will go a long way in reducing losses. Gray cast iron is ordinarily used for the melting pot and nozzles ("goose-necks"), but various special alloys have also been employed for the latter. Coatings are sometimes used on melting pots and nozzles, and alundum has been employed principally in order to prevent iron dissolution. The rate of production of die castings is, of course, much greater than in the case of sand castings, and the speed attained is dependent upon the size and type of the casting being made, the composition of the alloy cast, and the type of die that must be employed. When the castings are small in size and simple in shape, it is often possible to gate a number of them in one die; thus, six or more castings may often be die-cast at once. The rate of production of aluminium-alloy die castings varies from say 300 to several thousand pieces per day, depending upon the type of casting and the kind of machine used. In ordinary practice, die castings are given a rough inspection after removal from the die, and at the end of a run, the machine operator goes over the work and discards all wasters. Irrespective of how tightly the die halves fit or how carefully the dies are made, there is always some trimming to be done on the castings because of crevices placed in the die for venting purposes, or which exist accidentally owing to improper fitting of the parts. The small fins on the castings are trimmed by hand operators in the cleaning department, and usually these fins are scraped off with a scraping knife. If large, the fins are filed off, and gates are cut off by band saw or sprue cutter where necessary. Aluminium-alloy die castings are frequently finished in the highly polished condition, and ordinary buffing operations are employed in order to impart a high polish to the castings. At times nickel-plated aluminium-alloy die castings are specified, and these are produced, although the nickel plate is not always satisfactory.

Die-Casting Machines in General.—There are a number of different die-casting machines in use, and a great many have been patented. The fundamental principles upon which die-casting machines operate are all the same, involving, as they do, a melting pot for holding the alloy and a burner for heating it; a method for forcing the liquid alloy into the dies; and the die itself, together with mechanical devices for opening and closing it, and devices for ejecting the frozen casting. Roughly, die-casting machines may be divided into three classes, viz., (1) hand-operated; (2) semi-automatic; and (3) automatic. The processes in use may be divided into two classes: (1) pneumatic processes; and (2) plunger processes. The latter are now employed successfully in the United States, Canada, Great Britain, Germany, and elsewhere for the production of die castings in tin-rich, lead-rich, and zinc-rich alloys; but for the production of aluminium-alloy parts, the plunger machine is not well adapted, because of the higher temperatures employed and the resultant expansion of the cast-iron cylinder and plunger, which results in difficulties in working. Semi-automatic, pneumatic or compressed-air machines are largely employed for the production of light aluminium-alloy die castings. Several good serviceable machines, such as the Kralund, Hakanson, Bungay, Doehler, Stewart, and Sandage types are, or have been, in actual use. Many of the machines patented are absolutely useless from the manufacturing point of view, since they are either too complicated or get out of adjustment too easily. The mode of operation of the Hakanson machine has been described previously, and this is typical of aluminium-alloy die-casting machines. In the Kralund * apparatus the principle of operation is similar to that of the Hakanson machine. A movable ladle dips into the liquid alloy in the melting pot so that a charge of the latter runs in; after this, the ladle is moved so that its mouth is aligned accurately with the die opening. Then, the alloy is forced into the die by compressed air. In the Sandage machine † a bottom-pour melting pot is employed, the liquid alloy being delivered to the die under compressed air. The mechanical devices used for opening and closing the dies vary a great deal among the different machines, and there are various methods employed for cutting the feeder. Some automatic die-casting machines have been developed for rapid production, but these machines are usually exceptionally complicated, and they are expensive to build and keep in operation. It should be pointed out, as has been done by Harriman,[23] that unlike much industrial machinery, such as machine tools, which may be purchased from manufacturers, die-casting machines are not ordinarily made for sale by equipment builders, but these machines are largely designed and built by die-casting companies. Much secrecy attaches to the design and building of such machines. Fig. 167 shows a

---
* U. S. Pat. No. 1,299,738, April 9, 1919.
† U. S. Pat. No. 1,156,557, Oct. 12, 1915.

vertical automatic die-casting machine for aluminium-alloy work. Larger and larger die-casting machines are being made

Fig. 167.—*Vertical automatic die-casting machine (Harriman).*

yearly, and the largest die ever attempted was under construction in 1923; this weighs 8,000 lbs. with its ejector mechanism.

Recently, there has been a tendency toward the use of automatic machines for small castings.

The application of compressed air to die-casting machines for forcing the liquid alloy into the die has been of especial interest to die-casting manufacturers because of its ease of operation and simplicity, the positive action secured, and the fact that operating difficulties, such as the sticking of the plunger to the cylinder in plunger machines, are overcome. Of course, the principal difficulty in the use of air is that oxygen combines readily with metals at high temperatures, and when air is forced against the surface of a liquid metal oxidation is marked, and air is taken into solution and also entrapped in the metal. Many attempts have been made in specially constructed machines to prevent the direct impingement of the air upon the liquid alloy, but these have not been successful. However, by developments in the design of machines and methods for introducing the air, compressed air is now employed for forcing the alloy into the die in aluminium-alloy work. In the past, much difficulty was experienced in the use of compressed-air machines, and this has been eliminated largely by the use of dry air. The air is delivered to the machine in some installations after having been passed through heated coils. The aluminium-alloy machines are usually of the horizontal type, but there are, however, some large machines which are vertical and operated throughout with compressed air. The pressure used for forcing the liquid alloy into the dies ranges from about 100 to 1,500 lb. per sq. in., depending upon the kind of alloy cast and the size and shape of the casting. The more intricate the casting the higher must be the pressure, so that all parts of the mold may be properly filled. The size and section thus determine largely the pressure to be used. The lower pressures are employed for large and heavy castings. For ordinary aluminium-alloy castings the pressure is from 100 to 500 lb. per sq. in., with an average of 300 lb. for small castings. The devices which are employed for holding the dies together vary a great deal among different types of machines. It is important that the die be entirely tight, especially when the alloy is forced in under air pressure, since otherwise the liquid alloy would be forced out through any slight opening.

Brief reference has been made to the production of castings

by mechanical pressure in metallic molds, i.e., the Cothias process, and in passing it is of interest here to describe this process. The mold is made of cast iron, and the cores may be made of either steel or sand, steel being normally used. The mold is made in two separate units corresponding to the cope and drag of a sand mold, and the top half contains the cores. The cavity of the lower half of the mold conforms in shape to the desired outside shape of the casting. In production, the lower half is first heated to a suitable temperature for running the work, and then the exact amount of liquid alloy required to make the casting is poured into this half. The upper half of the mold (containing the cores and conforming to the inside shape desired in the casting) is then forced down mechanically into the lower half in the manner of a stamping press. The liquid alloy is thus compressed in the two halves of the mold, and is forced into all corners and intricacies of the mold cavity. This process is employed for the production of automotive castings in England, but it has not been used to any important extent in the United States.

Design and Preparation of Dies.—The production of a die is ordinarily expensive, and some dies may cost up to $1,500 or more, and require the work of a skilled mechanic for several months for completion. The design of die castings and dies, and the preparation of dies, requires high technical skill.

Die-casting dies consist of two main parts, which correspond to the cope and drag of a sand mold. Each half of a die may be made of a large number of component members, but when assembled, the dies must open and close in two sections. The internal cavities, holes, and other irregularities of a casting are produced by steel cores, and these cores are ordinarily made of the same steel as the die proper, e.g., chrome-vanadium steel, but tungsten-steel cores are also employed. A complete die may be described as a complicated mold which is made of steel, and is designed to be used many times for the same casting. Die-casting dies have been discussed by Pack,[22] among others, and it is not possible to treat the subject in detail in the present book. In a general way it may be said that in the design of dies the elimination of sharp corners is very important, because of the rapid chilling action when the alloy enters the mold. Thin, flat surfaces are avoided where possible, but if necessary they are ribbed, particularly if the casting is large; ribs serve a dual purpose in a die casting of this kind because they not only act as feeders in filling the mold, but also prevent the casting from becoming distorted and warped on handling while hot.

Ordinarily, one-half of a die is movable and the other half is stationary; the former is called the ejector die and the latter the cover die. The liquid alloy is injected through the cover die. The cores are actuated by pinions which engage in rack teeth on the cores, but levers may also be employed. It is generally preferable to place cores in the ejector die, which is drawn in a direction opposite to the melting pot of the machine, but at times it is better to place them in the cover die, owing to conditions which affect the method of parting. The die must be so constructed as to allow it to be parted and the casting removed quickly and easily, and dies are provided with ejector pins which are pushed into the die after each cast, thereby ejecting the casting. The ejector pins are carried in an ejector plate which is placed in back of the ejector die. The ejector pins pass through the die member to the die cavity. In order to remove the casting from the gate issuing from the cover die, a sprue cutter is employed, and this is usually simply a steel stud which projects from the ejector die far enough to enter the bushing in the cover die. A typical sprue cutter for castings having a hole in the center consists simply of a rod which is pushed through the center of the casting. After the alloy is forced into the mold, the sprue cutter is pushed through it in order to separate the casting from the frozen alloy of the feeder. Since the die is normally made in two parts, there must be a line of parting on the casting, and this parting line should always be placed at such a position as will enable easy ejection of the casting from the die. The position of the parting line is of great importance, and its determination requires detailed experience in the design of die castings. Of course, consideration should be given to the effect which the parting joint will have on the appearance of the finished casting, but if the die is constructed correctly, the line of parting will not be noticeable. Where practicable, the parting line may well lie on some edge of the casting.

The allowance to be made for contraction of the aluminium alloys in die casting should be known with accuracy, for use in design, and since the alloys are not free to contract normally, the contraction allowance must be arrived at by experience. The contraction allowance ordinarily used for the usual alloys is 0.004 to 0.005 in. per inch. In order to prevent the pocketing or entrapping of air in the die, it is necessary to vent, and vents are made simply by milling shallow crevices, 0.003 to 0.005 in. deep and $\frac{1}{4}$ to $\frac{1}{2}$ in. wide, across the die face from the cavity to

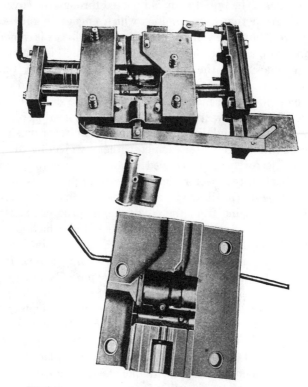

FIG. 168.—*Die-casting die and carburetor body produced by casting therein (Harriman)*.

the outside edges of the block. Of course, the machining and fitting of the various parts of a die constitutes one of the most important and difficult steps in the process, and it requires excellent workmanship. A poorly fitted die will operate with difficulty and also yield poor castings. The preparation of some typical dies has been described.[20] Fig. 168 shows the two halves of a die-casting die and a carburetor body produced thereby.

Production of Die Castings

In passing, it is of interest to mention the question of casting inserts in die castings, because the proper use of inserts has greatly broadened the field of application for these castings. Inserts may be defined as metal parts, ordinarily made of brass, bronze, iron or steel, which supply certain properties that die castings may lack. Inserts are of use also in providing more efficient means for the assembly of die-cast parts. The application of inserts may be divided into four general classes, viz., (1) to provide electrical properties; (2) to provide mechanical properties, such as greater durability or strength; (3) to provide economical and efficient assemblies; and (4) to provide lubricating facilities. Inserts must be anchored properly in the casting so that they will not pull out, turn, or move, and there are a variety of methods employed in practice for suitably holding inserts in place.

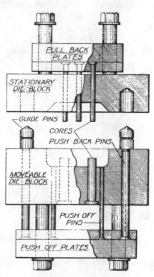

Fig. 169.—*Side view of a typical die (Harriman).*

Operation of Dies.—The operation of a typical die-casting machine has already been described, and here it is of interest to take up briefly the operation of dies in the process of die casting. Fig. 169 shows [23] a typical die which has cores in both halves, and Fig. 170 illustrates the cycle of operation. In a of Fig. 170, the die is shown in the closed position, and the cores are in place ready for injection of the liquid alloy. At b the pull-back plates are drawn back, and the die is open; the casting is shown in the movable block. At c, the casting is knocked out. At d, the die is in the operation of being closed; the push-back pins are bearing on the stationary block of the die in order to bring the push-off pins flush with the die cavity when the die is closed.

Figs. 171 to 174, inclusive, show views of a die used at the Light Manufacturing & Foundry Co., Pottstown, Pa., for the production of a beveled pinion made in two halves. The casting is so designed that the two sections fit accurately together and

604   *Metallurgy of Aluminium and Aluminium Alloys*

portions of each piece overlap the other piece so that when placed on a shaft and connected with a pin, the two pieces are permanently joined. Two views of the ejector half of the die are shown in Figs. 172 and 173, and the cover half is shown in Figs. 171 and 174. The two loose sections or heads which may be seen at the lower corners of the ejector die in Fig. 173 form two holes in each casting. When the die is open, these heads are held away from the main section of the ejector die by springs. When the die is being closed, pins on the heads slide into guide slots of the cover die, thus forcing the heads toward the main section

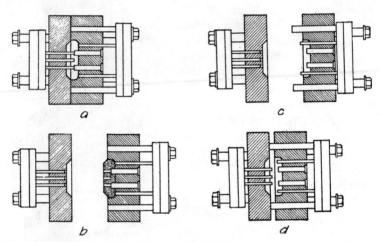

FIG. 170.—*Showing cycle of operation of a die in the production of a casting (Harriman).*

and carrying the pins into position for forming the holes. One of the guide slots in the cover die is shown in Fig. 174. Referring to Fig. 171, the large number of screws in the back of the cover die is employed for holding the different pieces of steel together which are used in making up the block. The four larger holes shown in this figure are guide holes for the pins which cause the two members of the die to fit together accurately when the die is closed. The two smaller holes are gates through which the liquid alloy is injected into the mold. A side view of the ejector half of the die in Fig. 172 shows the plate which is firmly connected to the face of the die cavity by means of a number of small round rods (ejectors). When the plate is moved by hand lever toward the front of the die, the ejectors push the casting out of the die cavity. The air vents can not be seen in the figures, but these are usually placed on the face of the die, and are so small that the liquid alloy will not be forced through them.

Materials Used for Dies.—For the manufacture of die castings in zinc-rich, tin-rich, and lead-rich alloys, dies made from simple low-carbon steel are entirely satisfactory, but the problem of securing a suitable material for dies to be used in the production of aluminium-alloy die castings has been a serious one. A wide variety of steels has been tried for die making for aluminium-

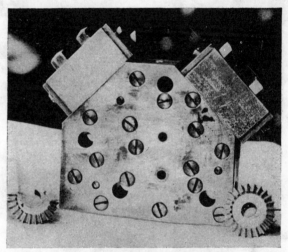

FIG. 171.—*Rear view of cover half of a die for a beveled pinion* (*The Foundry*).

alloy castings, and alloy steels of the chrome-vanadium type have proved to be reasonably suitable. The dissolution of iron by aluminium from cast-iron melting pots is, of course, well known, and for die casting in these alloys it is necessary that the solvent action be low. The presence of iron in amounts up to 2 per cent in No. 12 alloy is not serious from the point of view of the die casting itself, but when iron from the surface of a die is dissolved, the die necessarily becomes imperfect, and accuracy and uniformity in the product can not obtain. It is for the latter reason that the solvent action of aluminium alloys upon the die material should be very low, and a cardinal principle in die-casting practice for any alloy is that the metallic mold shall be capable of withstanding the action of the liquid alloy. Special alloy steels of various compositions are favored for dies, and ordinary chrome-vanadium steels are preferred. Some special die-casting steels are on the market. One difficulty with ordinary

FIG. 172.—*Side view of the ejector die for a beveled pinion; cf. Fig. 173 (The Foundry).*

steels, and with alloy steels as well, for dies, is that checking or cracking of the surface occurs in use, and when this happens the die is useless and must be discarded. With ordinary carbon steel, the dies crack and check after 500 casts with aluminium alloys, and they will last only about 1,000 to 2,000 heats with chrome-vanadium steel in the unheat-treated condition. When properly heat-treated, chrome-vanadium steel dies will last from 10,000 to 15,000 casts.

The actual die faces for aluminium-alloy castings are usually made of ordinary chrome-vanadium steel or special chrome-vanadium or other alloy steels, and particular steels have been

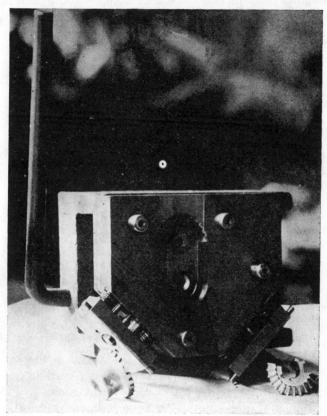

FIG. 173.—*Front view of the ejector die for a beveled pinion; cf. Fig. 172 (The Foundry).*

developed which are known in the steel trade as "die-casting steel." The heavy bases and frames of the dies are ordinarily made of cast iron, while the cores are frequently made of tungsten steel of ordinary tool grade. There is a great deal of heat-treating to be done in the preparation of dies, particularly for the purpose first of softening the steel so that it can be readily machined, and second for again hardening the dies. A chrome-vanadium steel used for the production of dies has the following approximate composition: chromium, 2.10 per cent; vanadium, 0.35; carbon, 0.40; manganese, 0.65; and silicon, 0.10 per cent. The question of the performance of dies in aluminium-alloy die casting has been discussed at some length by Tour,[27] and he has directed attention especially to the cracking of dies owing to thermal fatigue. For aluminium alloys, the heat absorption value may be taken as about 562 calories (i.e., heat removed per unit volume, cals. per c.c.), which is substantially higher

FIG. 174.—*Cover half of die for a beveled pinion* (*The Foundry*).

than for lead-rich, tin-rich, and zinc-rich alloys. Failure of dies is often brought about by surface checking and cracking, and the mechanism of the action which causes heat cracks in a die is very similar to that causing cracks in ordnance because of hot gas. Fig. 175 shows an enlarged portion of the surface of an actual die block after the production of about 20,000 castings. The heat cracks in the die naturally give rise to a network of small fins, imparting thereby a rough appearance to the surface of the casting. A die can be used until these fins become of such size or thickness that they can not be economically removed from the castings, or until they interfere with the operation of the die. The mechanism of the action which results in heat cracks can not be discussed in detail here, but reference may be made to the paper by Tour [27] for detailed information. Briefly, however, the cracking is the result of alternating thermal fatigue when the alternation of stresses is above the endurance limit.

Aluminium Alloys for Die Casting.—At the present time, die castings are made in (1) zinc-rich alloys; (2) tin-rich alloys; (3) lead-rich alloys; and (4) aluminium alloys. The nominal composition employed for high zinc alloys is 88 : 8 : 4 zinc-tin-copper; that of the high tin alloy is 86 : 6 : 8 tin-copper-antimony; while that of the high lead alloy is 85 : 15 lead-antimony. A great variety of alloys have been experimented with for die casting, but it has been found in the case of the zinc-

FIG. 175.—*Appearance of surface of die after about 20,000 castings (Tour).*

rich, tin-rich, and lead-rich alloys that comparatively few alloys meet all trade requirements. As to aluminium alloys for die casting, the situation is somewhat different. In the past, much experimentation was carried out with various kinds of light aluminium alloys, and finally the ordinary 92 : 8 aluminium-copper alloy was standardized as the preferable alloy for this work. For several years this was the only alloy employed to any extent, but during the past five years a number of special alloys have been employed. These include principally the silicon-containing aluminium alloys. For die casting in the case of binary aluminium-copper alloys, the copper content is usually 6 to 14 per cent, depending upon the type of casting made. The 90 : 10 aluminium-copper alloy is favored considerably now, as is the alloy of the nominal composition 90 : 8 : 2 aluminium-copper-nickel.

The usual investigations which have been made on the properties of metallic alloys have been confined in the main to sand-cast, chill-cast, wrought, and annealed alloys, and but few studies have been made of die-casting alloys—particularly in aluminium alloys. Just what constitutes a good aluminium die-casting alloy is a matter of considerable conjecture, but practical experience indicates that at least three items are highly important; these are: (1) the melting point of the alloy; (2) its solvent action upon iron; and (3) its strength and elongation at elevated temperatures. The zinc-bearing aluminium alloys are ruled out for die-casting work because they rapidly attack iron and steel when liquid, and further because they are extremely hot-short. Alloys having fairly high elongation in the solid state at elevated temperatures are desirable because of the inherent conditions involved in the die-casting process. A metallic core, unlike a green-sand core in sand-casting practice, is not compressible, and when an alloy forced into a die starts to solidify, which it does rapidly, the normal contraction of volume takes place. That is to say, the alloy contracts in the die so far as possible, but it is prevented from normally contracting with the result that stresses are set up; if the elongation of the alloy is low at high temperatures, the alloy will crack in the die. Low contraction in volume is also desirable in die-casting alloys, and the silicon-bearing alloys are especially good in this respect. Other than the aluminium-copper alloys mentioned above, certain aluminium-copper-nickel alloys have been employed for die casting, and in these the copper content is 7 to 9 per cent and the nickel 1 to 3 per cent. Certain ternary aluminium-copper-silicon alloys containing up to 12 per cent silicon and up to 5 per cent copper, e.g., 87 : 3 : 10 aluminium-copper-silicon, are now being used. Interesting aluminium-nickel-silicon alloys have also been recently developed for die casting. Pack [*] states that a suitable die-casting alloy contains 85 per cent aluminium, 3 to 6 copper, 3 to 6 nickel, and 1.5 to 4 per cent silicon.

Properties of Die Castings.—The tensile properties of aluminium alloys, when die-cast, may be comparable to the results obtained from casting such alloys in a chill mold. The tensile strength and elongation are both higher than for similar alloys

[*] U. S. Pat. No. 1,449,052, March 20, 1923.

cast in sand, and for the light alloys the specific gravity is slightly greater. Die castings are more dense than sand castings, in the general meaning of the term, but they may not necessarily be free from blowholes and porosity, despite the extravagant claims often made for them in regard to this matter. As a rule, reputable die-casting makers are loath to guarantee specific tensile properties in a die casting. The soundness of castings may be readily determined by using a perfect casting as a standard, and weighing subsequent castings against it. This method will readily detect the presence of small blowholes. With specific regard to the soundness

FIG. 176.—*Blowholes in die-cast 92 : 8 aluminium-copper alloy; etched NaOH; oblique illumination;* × 2.5.

of die castings, it must be admitted that such castings produced by forcing the alloys into the dies by compressed air are generally unsound, and castings made in plunger-type machines are also un-

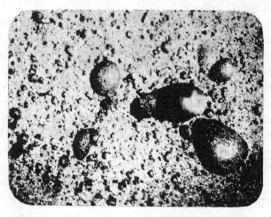

FIG. 177.—*Blowholes in die-cast 92 : 8 aluminium-copper alloy; etched NaOH; oblique illumination;* × 2.5.

sound. Blowholes and related defects in die castings are the result largely of entrapped air, and it is practically impossible to avoid the occurrence of entrapped air in a die casting made by the ordinary methods. Figs. 176 and 177 show blowholes of a typical kind; Fig. 176 shows the unsoundness encountered in a small shank section of a casting made in 92 : 8 aluminium-copper alloy, while Fig. 177 shows the appearance of large holes in another section of the same casting. While the surface shells of die castings are sound and have small grain size owing to the chilling effect of the mold, the interior is invariably porous and blowholey, and die castings should not be used, for important construction parts which must withstand stresses.

The tensile strength of 92 : 8 aluminium-copper alloy, when die-cast, is about 21,000 to 23,000 lb. per sq. in., and the elongation is 1 to 1.5 per cent on a 2-in. length. The brinell hardness is about 60. It has been suggested that die-cast parts in aluminium alloys, which are susceptible to heat treatment, should be heat-treated in order to enhance the physical properties, but this is evidently quite out of the question. When a die casting is heated in a furnace for annealing or quenching, the gas in the internal holes expands and causes bad blisters on the surface, so that the heat treatment of die castings seems to be rather largely precluded, unless the parts are entirely refinished after such heat treatment.

Uses of Die Castings.—The general uses of aluminium-alloy die castings have already been indicated in Chapter VII, and Figs. 77 and 78 of that chapter show some typical die-cast parts. Aluminium-alloy die castings are admirably adapted to a variety of uses, particularly small parts in automotive work, but they are employed widely in the engineering trades, and their specific applications are too many to enumerate. In motor-car construction the following parts are die-cast (often in 92 : 8 aluminium-copper alloy or alloys richer in copper) and used on a representative number of the leading makes: control sets; starting, ignition and lighting system parts: speedometer parts; magneto parts; small housings; bushings and nuts; gears to transmit motion; carburetor covers; bearing backs; water-pump impellers; fan pulleys; carburetor bodies and carburetor throttle plates and levers; magneto end plates, breaker boxes, and switch levers. Motor-cycle crankcase parts are also die cast in aluminium alloys. The manufacture of aluminium-alloy pistons for automobile and aircraft engines has been largely by the permanent-mold process, but partly by die casting and sand casting. Die casting is not suitable for piston production because of the extraordinary porosity and unsoundness found in die castings.

Other uses of aluminium-alloy die castings include the following: cases and covers for electric mine lamps; oil bowls for mine flame safety lamps; brush holders; door checks; camera frames; adding-machine parts; unions; phonograph and player-piano parts; motor housings and covers; clock bezels; stamp affixers and check protector parts; coin-changing devices; and various small parts for scientific and testing instruments, and for machinery. Some uses to which aluminium-alloy die castings were put during the War include the following: parts for gas masks, such as breather tubes; small parts for the Lewis machine-gun, and for the Browning machine-gun; navy and army binocular housings; speedometer parts, steering-wheel accessories, gasoline-regulating devices, carburetors and carburetor parts, ignition housings, and the like, for trucks, tanks, and aircraft; signal pistols; parts for submersible bombs; parts for rifle and hand grenades; plugs for trench-mortar shells; parts for aircraft bombs; and surgical appliances.

The progress of the aluminium-alloy die-casting industry most recently has been in the direction of increase in the intricacy of design and increase in the size and weight of castings made. Aluminium-alloy die castings are finding increased application in the automotive industry, and a much wider use than formerly in household equipment such as vacuum cleaners, washing machines, and electrical appliances. The development of radio broadcasting and receiving has resulted in the production of hundreds of thousands of die castings of various sorts for radio sets. In connection with the use of aluminium-alloy die castings for automotive work, it should be especially emphasized that certain applications of such castings, i.e., where strength and soundness are important, have been made incorrectly from the engineering point of view, and it is certainly hazardous to entrust life in an important part to a die casting. Permanent-mold castings should be employed for certain parts where die castings are now used.

PRODUCTION OF PERMANENT-MOLD CASTINGS.

The commercial production of aluminium-alloy parts by permanent-mold casting is a relatively new process, although, as mentioned, the casting of metals and alloys in permanent types of molds dates back to the early stone and bronze ages. Crude burnt clay molds were made by prehistoric man for casting rude implements of warfare, and later open-stone molds were employed for the same purpose. Bronze molds for casting gouges, socked celts, and other tools were still later used, and good examples of old stone and bronze molds are preserved in museums. Molds of the permanent type have been employed in recent years for the production of castings in both ferrous and non-ferrous alloys, and the general principles and history of the process have been outlined by the author and M. E. Boyd.[42] In the light alloy field, permanent-molds have been used largely for casting pistons for internal-combustion engines, and this development dates from about 1912–1913, when the aluminium-alloy piston was introduced for racing motor cars. The production of aluminium-alloy pistons by permanent-mold casting has also been discussed by the author and M. E. Boyd.[43] In the past five years, with the increasing demand for aluminium-alloy castings, especially in the automotive industry, permanent-mold castings are becoming important, and they have displaced sand castings for a number of purposes. There are several advantages in making aluminium-alloy castings in permanent molds, including the better mechanical properties gained owing to the rapid chilling effect of the mold and the increased rate of production which lowers costs. Permanent-mold castings also have a good surface appearance, which is of importance when the castings are not machined or buffed. Of course, in the case of small parts which do not need to be particularly sound or strong, the permanent-mold process can not possibly compete with die casting; but for making medium-sized castings such as pistons, small housings, and other parts which are not too large and which must have good physical properties, this process presents distinct advantages.

In the discussion given above on die casting, it was pointed out that permanent-mold castings are often called " die castings," but by definition the term permanent-mold casting is used to connote the product obtained by pouring liquid alloys into metallic molds, the alloy going into the mold under the force of gravity solely. The permanent-mold process is largely confined to the production of castings in which metallic cores can be used, but some castings have been made using sand cores; however, the process is best adapted to the use of all-metallic molds and all-metallic cores. The essential features and properties of permanent-mold castings are similar to those of otherwise identical die castings, except that the former castings are sound, whereas the latter are not. In contradistinction to die casting, heavy aluminium alloys, i.e., aluminium bronze, are cast by the permanent-mold process. Taken by and large, permanent-mold castings in aluminium alloys are fairly sound as contrasted with die castings, and the most usual defects encountered are shrinkage holes, blowholes, cracks, holes due to entrapped air, and overlaps due to the meeting of liquid alloy in its flow in the mold. The more theoretical aspects of permanent-mold castings are interesting, but they can not be discussed in detail here.

While practically no machining is done in the case of die-cast parts, some considerable machining may be done on parts cast in permanent molds. For automotive hardware, handles, and other small parts, little machining is ordinarily required, but in the case of pistons considerable machining may be necessary. The amount of machining depends upon the type of casting and its design and the method of molding. Actually, the permanent-mold process appears to be best adapted for the production of castings like pistons, and it does not seem to be good for the production of small parts which can be made by die casting, unless such parts must be comparatively free from blowholes. The permanent-mold process may be properly regarded as standing midway between sand casting and die casting, and it has a definite field not covered by either. For large castings and complicated small castings the sand-casting method should be employed. For simple and small castings, and even fairly complicated castings which do not need to be particularly strong nor sound, the die-casting method is good. In the case of reasonably simple and small castings, which are to be made in large numbers and which must be strong and sound, the permanent-mold process is advisable. There is little question but that many castings are made in the sand foundry which might profitably be made by die casting, and it is also true that some castings are made in permanent molds which should be made by die casting. On the other hand, some parts are made by die casting which should be produced in the sand foundry or by permanent-mold casting, and it is only possible to determine the correct method to employ by consideration of the castings to be produced.

Permanent-mold casting is necessarily a quantity-production process, and not many parts should be considered practical for casting in permanent molds in less than lots of 1,000. This is so because of the expense of making the molds, although the mold cost in permanent-mold work are not nearly so high ordinarily as die cost in die casting. Almost any kind of casting can be made in a permanent mold, but, as explained previously, the process is applied principally to pistons in the case of the light aluminium alloys, although various other kinds of castings are made, and aluminium-bronze castings of various types are poured in permanent molds. The design of the mold is a most important factor, and so far experience is the best guide, since few actual technical investigations have been carried out on the subject. The advantages of making castings in permanent molds are comparable to making them by die casting. Castings made in permanent molds are accurate in size and uniform as to successive casts; machining costs are either largely eliminated or much reduced over sand casting; and the output is greater than can be obtained by sand casting. Of course, the output by casting in permanent molds is not so rapid as by die casting, and generally the accuracy is not as great. So far, the permanent-mold process has neither been well understood nor appreciated either in the foundry or by consumers of castings, but the author looks for great expansion in this method of casting.

Mode of Manufacture of Permanent-Mold Castings.—The process of making castings in permanent molds consists essentially in pouring a liquid alloy into a previously heated and assembled mold, disassembling the mold as soon as the alloy has frozen, and removing the casting. The operations consist in assembling the mold, heating it to the proper temperature (in practice the mold is kept hot by burners), inserting the cores, and pouring the alloy. The cores are removed in the proper order and at a suitable interval of time after the alloy has been poured, and the mold is then taken apart so that the casting can be removed. The resultant casting has a smooth finished surface, and it may require no machining other than the removal of gates and fins and trimming. However, the amount of machining to be done depends upon the type of casting as explained above. The principles of producing permanent-mold castings have been treated in detail in the literature by Johnson,[31] an anonymous writer,[35] and by the present author and M. E. Boyd.[42, 43] As in the case of die casting, the following three items are essential for the successful production of permanent-mold castings: (1) a metallic mold which will withstand the action of the liquid alloy, as well as rapid changes of temperature; (2) an alloy suitable for the process; and (3) a design of gating and a method of pouring that will ensure a sound casting and one which fills the mold completely. As in the case of both sand casting and die casting, the permanent-mold process has its limitations under modern conditions, and while it is possible to cast practically any part in a permanent mold, it is not feasible to attempt to do so. Of course, the molds must be made so that their cost of operation and upkeep shall be small, and hence permanent molds are made as simple as possible. Usually, two men are required to operate a permanent mold, but lately so-called one-man molds have been built which reduce operating costs. Some castings are made in permanent molds with collapsible cores, but not many such castings are produced.

The production of aluminium-alloy pistons in permanent molds will be taken up fairly fully in later paragraphs, and here it is of interest to consider the general principles of the permanent-mold process. Székely* has patented the general principles of permanent-mold casting, but many improvements have been made in the past 10 years particularly in the construction and operation of molds for the production of particular kinds of castings. It should be pointed out here that the development and use of the permanent-mold process, as applied to aluminium alloys, has been much hampered by litigation, and that many different types of molds have been patented for the production of different kinds of castings. Referring to the Székely patent Figs. 178 to 181, inclusive, show a form of mold. Fig. 178 is a plan of the closed mold, which is made up of seven parts; Fig. 179 is an end view of the mold; Fig. 180 is a front elevation as

---
* U. S. Pat. No. 841,279, Jan. 15, 1907.

seen from the left in Fig. 179; and Fig. 181 is a plan showing the mold thrown open. Referring to the figures, C, in Fig. 181 designates the casting produced in the mold. The mold itself

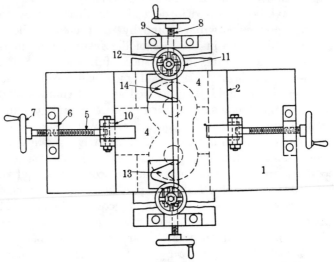

FIG. 178.—*Plan of permanent mold, closed (Székely)*.

consists of a base *1*, movable sides *2*, movable ends *3*, and movable top plates *4*. The mold sides *2* are slidable on the base and are moved thereon by screws *5*, rotatable in nut bearings *6* on

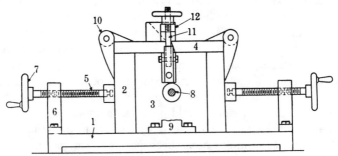

FIG. 179.—*End view of permanent mold shown in Fig. 178 (Székely)*.

the base and having crankwheels *7*. The mold ends *3* are also slidable on the base and are operated by screws *8*, rotating in nut bearings *9* on the base. The top plates *4* are hinged at *10* to the respective sides *2* and turn about these hinges, which connect the top plates to the respective mold sides *2*. When the top plates

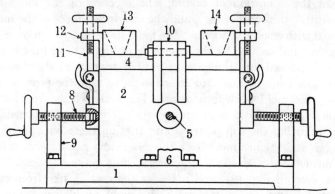

FIG. 180.—*Front elevation of permanent mold as seen from the left of Fig. 179* (Székely).

are down in place, they are locked by means of swing bolts *11*, which are hinged to the respective mold ends *3* and engage

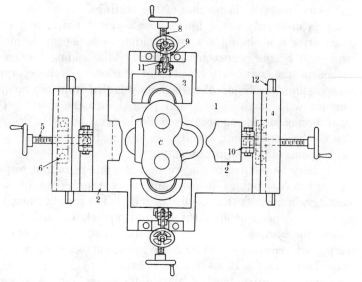

FIG. 181.—*Plan of permanent mold, thrown open; refer to Fig. 178* (Székely).

slotted lugs *12* on one of the plates *4*. The swing bolts have a wheel nut *11a*. In one of the top plates, there is an outlet *14* for the gases from the mold. Suitable core recesses are provided in the mold to suit the particular object to be cast.

In the operation of casting, the sections of the mold are separated, the cores are set, and the interior surfaces of the mold coated with a suitable wash. The mold sections are then brought together and secured by the screws, as in Fig. 178, and the liquid alloy is poured into the gate *13* until it appears at the riser *14*. After the alloy has time to freeze so as to take the form of the mold, the mold is thrown open. The time which the casting remains in the mold is about 20 secs. The mold is opened by first releasing the top plates *4* and turning them back on their hinges, and then running back the mold sides *2* and mold ends *3* by means of their respective screws. This is the position of the parts seen in Fig. 181, and it leaves the casting free from confining pressure or contact with the mold on all sides and at the top.

The essential steps in the production of finished permanent-mold castings are these: (1) design of the casting and mold; (2) set-up of the mold; (3) melting and pouring the alloy; and (4) machining and inspection of the castings. In melting, the alloy is ordinarily made up in an alloying furnace, and then transferred while liquid to a small holding furnace near the molds and kept at the proper temperature. The holding furnace is either of the stationary-crucible or stationary iron-pot type. For casting, the liquid alloy is simply ladled from the holding furnace with small dippers and poured into the molds. The operation of molds has been discussed in detail above in describing the Székely patent, and it is of interest to discuss some general aspects of the subject broadly.

It may be pointed out again that the rate of output of permanent-mold castings is generally more rapid than of sand castings, but slower than of die castings. The speed attained is, of course, dependent upon the size and type of the casting being made, the composition of the alloy used, and the type of mold employed, referring more especially to the number of cores which must be inserted. Brass or steel inserts are not often placed in permanent-mold castings of aluminium alloys, but when used, the necessity for placing them slows down the rate of production. The output with one-man molds is normally greater than with the ordinary type, but this may not always be so, although the former mold reduces labor costs of operation. Molds may be difficult of operation because of the draft or of many cores which must be placed before each casting is poured, and the removal of castings always takes longer than from a die in die casting. In many types of permanent molds, no ejector pins are used for ejecting the casting, and the casting may stick to the mold, particularly if the mold surface is rough. Some small and simple castings may be gated together in one mold so that several may be cast together at once as in a sand mold. In the case of piston practice, the rate of production in ordinary molds is from 200 to 300 pistons for two men in 9 hours, depending upon conditions.

The alloys used for aluminium-alloy permanent-mold casting are usually the same as those employed in die casting. The question of mold and core temperatures in permanent-mold casting practice is exceedingly important. In practice, molds are operated at 325 to 525° C. for piston production, and the cores may be kept somewhat cooler. Molds are ordinarily brought up to temperature by heating with burners, and the flame is shut off when pouring is started. Some molds are designed with inserts of greater heat conductivity than the mold proper in order to cause more rapid transfer of heat, and thus equalize the rate of solidification of the alloy. Washes and coatings for cores and the inside surfaces of mold cavities are practically useless, and need not be applied. However, in practice, a number of different dressings and coatings are used, including linseed oil, tallow, and a mixture of sodium silicate and lime made into a thin paint with water. In ordinary practice, permanent-mold castings are given a rough inspection by the molder after removal from the mold, but another inspection is given after finishing. After cooling on removal from the molds, the castings are freed of gates and risers by cutting with a band saw, and are then given a rough cleaning by grinding on a wheel. In the case of pistons, a great deal of machine work is required, but for simple castings the parts may be finished by polishing after the removal of gates and fins.

Permanent Molds in General.—The three requisites for successful permanent-mold castings are, (1) a suitable mold from the point of view of the material employed for the body of the mold and for the cores and from the point of view of design; (2) a suitable alloy; and (3) a suitable design of casting. Since a different type of mold is employed for every type of casting produced, it is quite useless to attempt to describe many types, but the fundamental principles upon which molds are built and operate are all the same. These principles involve the mold itself, together with its cores; the rationale of assembling the mold and pouring the alloy; and the furnace for heating the alloy. As has been mentioned, the alloy enters the mold simply under the force of gravity, and is put in by ordinary pouring as in the case of sand practice. In one-man molds, various mechanical devices are employed for opening and closing the mold and ejecting the castings, but in both ordinary molds and in one-man molds the cores are set by hand. The mold itself is heated over its exterior surface by small burners, but, while the question of mold temperatures is exceedingly important, devices for heating the molds are usually crude and do not ensure uniformity of temperature in the mold.

Numerous patents have been taken out pertaining to molds for the production of aluminium-alloy pistons in permanent molds, and a description of one of these typical devices * may be given. Figs. 182 and 183 show the cross-section and plan of a device for casting skirted pistons in light aluminium alloys for internal-combustion engines. Referring to the figures, the mold is made up of a base member a and complementary members b and c; the latter are movable on the base member a toward and from each other. The main core of the mold is indicated by d; this is formed of a number of sections suitably recessed to form the bosses and ribs on the interior of the pistons. Diametrically opposite core pins, e and e, are arranged to project into the mold cavity to form openings through the bosses 3 of the piston casting. These core pins are in the form of rods that are mounted in openings of the mold members b and c.

The mold members b and c are formed with complementary cavities or recesses b' and c', which together form the mold gate. The gate thus formed, as will be seen in Fig. 183, is disposed midway between the internal bosses 3, of the piston casting. The gate is constructed so that a liquid seal is formed, thus preventing the passage of air and gases into the mold cavity, and also reducing splashing of the metal within the gate during

---
* U. S. Pat. No. 1,296,592, March 4, 1919.

pouring. Also, the gate is preferably formed so as to lead into the mold cavity at the top and bottom and at intermediate

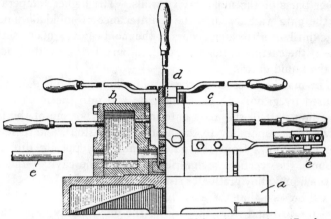

FIG. 182.—*Side elevation of permanent mold for piston castings (Bamberg).*

points between the top and bottom. Suitable means are provided for heating the walls of the mold members *b* and *c*. The core pins are kept preferably at a lower temperature than the

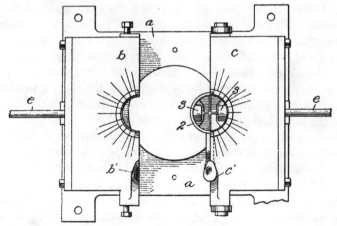

FIG. 183.—*Top plan of permanent mold for piston castings (Bamberg)*

other parts of the members *b* and *c*. When the liquid alloy is poured in (at a temperature several hundred degrees higher than that of the mold walls), it is subjected to a very strong chilling action at the points adjacent to the core pins e, e, and to a less marked, but still substantial, chilling action adjacent to other parts of the mold cavity walls. The higher temperature of the gate walls ensures the maintenance of liquid alloy in the gate until after freezing occurs in the mold cavity; hence shrinkage of the casting is taken care of, in part, by the feeding action of the liquid gates.

In making a casting, the liquid alloy, at about 750° C., is poured by gravity through the gate and into the mold cavity. The first part of the alloy entering fills the bottom of the mold cavity (corresponding to the head of the piston), and thereafter the alloy flows from the gate opening in opposite directions around the vertical cylindrical parts of the cavity—the level of the liquid alloy gradually rising until the cavity is full. Hence, it will be seen that the alloy freezes progressively from the bottom upward and from the side of the mold cavity opposite the gate in a direction toward the gate. In a short time after pouring, the core pins e, e are withdrawn, and then after another short interval of time the core d is removed, the mold members b and c are separated, and the casting is taken out. Substantially the same principles are described in other patents for the production of fuse bodies, rings,* and more especially of skirted pistons.† Methods of gating and coring are very variable in connection with the design of molds for particular castings, and details of these matters are without the scope of this work.

Since the problem of heat distribution in the mold is of great importance, as related to the quality of the casting produced, this must be taken into consideration when designing the mold. As the liquid alloy is poured into the mold, the heat is, of course, transmitted to the mold, causing it to become hotter with successive pours. If the mold became too hot, the operation would be slowed down considerably and the favorable effect of chilling would be lost. This makes it necessary to provide for proper cooling of the mold, and in practice this is usually taken care of by radiation rather than by water cooling. When making castings having thick and thin sections in contiguity, it is necessary to make some provision for having them freeze at about the same rate, since otherwise cracking will occur at the juncture of

---
* U. S. Pat. No. 1,323,938, Dec. 2, 1919.
† U. S. Pat. Nos. 1,296,596; 1,296,597; 1,296,589; 1,296,591; and 1,296,593, March 4, 1919; and 1,420,903, June 27, 1922.

the thin and thick sections. Attempts have been made in the direction of equalizing the rate of cooling by the use of inserts of greater heat conductivity in the mold which are placed adjacent to the heavy sections. Referring to the question of gating and the mode of introducing the liquid alloy into the mold, this is of the greatest importance, and many of the defects in permanent-mold castings can be traced to incorrect gating and uneven rise of the liquid alloy in the mold, as well as to agitation of the alloy. A wide divergence of opinion exists regarding gating methods, some designers preferring to adhere to simple gates, while others adopt weird appearing gates which are purported to give the desired flow of alloy upwards in the mold cavity.

Fig. 184 shows a permanent mold assembled for the production of dynamo-starter castings, with two of the castings in front of the mold. Fig. 185 shows the mold dissembled.

Design and Preparation of Molds.—While the preparation of a permanent mold is fairly expensive, machining costs are not so high as with dies for die castings, largely because the gray iron used for permanent molds is readily machineable, whereas the chrome-vanadium steel employed for dies is not. The general requirements of design, as applying to die castings, apply also in the design of permanent-mold castings. Molds, and the coring, should be so designed that the molds are readily parted and the cores easily removable. As stated previously, ejectors are not used in ordinary permanent molds, but they are applied in the construction of one-man semi-automatic molds. The use of ejector pins is advisable since these tend to increase production, and the application of pins to ordinary molds seems entirely warranted. The contraction allowance made for permanent-mold casting varies with the size and shape of the part, but for pistons cast in 92 : 8 aluminium-copper alloy is 0.10 per cent.

The pocketing of air in a permanent-mold casting is serious, and many gating devices are adopted to avoid this trouble. In the design of the gating, it should be remembered that the temperature of the mold is considerably lower than that of the liquid alloy, and that the alloy, upon coming in contact with the mold, forms a skin where it touches the mold and becomes more sluggish in its flow. This results in a more rapid rise of the liquid alloy through the heavy sections of the mold than through the light sections. On this basis, a mold filled entirely from a bot-

Fig. 184.—*Permanent mold for making dynamo-starter castings; assembled* (*Guillet*)

Fig. 185.—*Permanent mold for making dynamo-starter castings; dissembled* (*Guillet*).

tom gate is not very satsifactory, although molds of this type are employed. It must be obvious that gates which feed the bottom of the mold cavity first and feed the casting along a vertical wall are theoretically better at least, since such gating does not disturb the skin of semi-fluid alloy in the bottom of the mold. It is of importance to have the pouring gate of the correct angle with relation to the casting, and the size of the feeder, or so-called choke, is also important. If the gate is vertical or nearly so, there is danger of the liquid alloy splashing when it strikes the bottom, and this would be undesirable. On the other hand, if

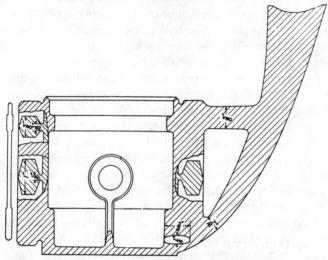

FIG. 186.—*Transverse section showing method of gating piston.*

the angle of the gate is very acute with reference to a horizontal line through the bottom of the mold, then there is a tendency for the liquid alloy, on being poured, to shoot up the opposite side of the mold cavity. A pouring gate with a moderate angle of inclination and a moderate feeder to the main body of the casting can be made quite readily after a few trials. Fig. 186 is a sketch showing the transverse section of a piston with pouring gate, feeders, and shrink pads attached.

In connection with gating methods, the size, shape, and position of the so-called shrink pads are of importance. There is little or no agreement as to how the shrink pads should be designed, and all manner of shapes in various positions with relation to the casting are in actual use. In the case of pistons, in some cases no pad is used on the side opposite the pouring gate, and apparently without harmful results. The casting of pistons without shrink pads on the side opposite the gate would be an economic procedure, since it would eliminate a large amount of scrap. While a shrink pad is used primarily for the purpose of feeding liquid alloy to the casting, just as a riser is used in sand practice, it has a secondary function, viz., it undoubtedly aids in regulating the flow of liquid alloy upwards, as to rate and uniformity. Coring practice in permanent-mold work is very variable depending upon the design of the casting, and 3-, 5-, and 7-piece cores are usual for hollow castings. The question of coring has been discussed by the author and M. E. Boyd [43] but can not be taken up in detail here.

The question of the preparation of dies for die casting has been covered, and the general principles of machine-shop practice, as applied to die making, also apply to the preparation of permanent molds. It should be emphasized, however, that if a good surface is given to the mold cavity, much less difficulty will be had than if the surfaces are rough and uneven.

Materials Used for Permanent Molds.—In making light aluminium-alloy castings by the permanent-mold process, the factors governing the selection of a suitable mold material are substantially the same as in die casting, but the conditions under which a permanent mold operates are less severe than those under which a die works. Gray cast iron is employed largely for making molds for permanent-mold casting, although a great many materials have been tried. The cores are made of tungsten steel of the usual tool-steel grade, and at some plants chrome-tungsten steel is used. Rix and Whitaker [33] recommend the use of close-grained cast iron. In such a mold, from 5,000 to 7,000 castings have been obtained. Molds made of cast iron high in graphitic carbon and low in combined carbon are preferred. A typical analysis for gray cast iron used for permanent molds is as follows: 0.14 per cent combined carbon, 3.35 graphitic carbon, 2.40 silicon, 0.43 manganese, 0.10 sulphur, and 1.3 per cent phosphorus. Another composition given is as follows: 0.89 per cent combined carbon, 2.76 graphitic carbon, 2.02 silicon, 0.29 manganese, 0.07 sulphur, and 0.89 per cent phosphorus.

High phosphorus is regarded as necessary to give fluidity to the iron and to obtain sharp molds when cast.

Production of Aluminium-Alloy Pistons.—Since the permanent-mold process has been used largely for the production of motor pistons, description of a typical plant in which such pistons are made may be found of interest. The alloy is made up in tilting iron-pot furnaces of an usual type, holding about 375 lb. of alloy, and for pouring, the liquid alloy is transferred to holding pots. Fig. 187 shows a view of a mold used for pistons; this is assembled ready for pouring, the cores being in place; the pre-heating burners for heating the mold are shown attached to the set-up. Fig. 188 is a front view of the same mold, showing the parting and lock, while Fig. 189 shows the open mold with the cores removed. The method of gating and the mold cavity may be seen in Fig. 189; the white on the gates is the lime-sodium silicate wash. Fig. 190 shows the cores of a piston mold assembled, indicating the position of the separate pieces; the two wrist-pin boss-hole cores are shown, one on either side of the assembled main cores.

In production, each piston mold is operated by two men, a molder and a helper. The molder dips the liquid alloy from the holding pot, pours it into the mold, and then returns the dipper to the top of the holding pot in order to keep it hot. This operation requires about 10 secs., depending upon the dexterity of the molder. The actual time period of pouring is $2\frac{1}{2}$ to 3 secs. While the molder is returning the dipper to the top of the pot, the helper removes the core pin or wrist-pin core on his side of the mold, and lifts out the center core section. Then, the helper lifts out the back core section, and the molder removes two side core sections as does the helper. The molder then unlocks the mold, and both men help to open the mold. Next, the molder removes the casting while the helper cleans off the mold, brushing out any chips or aluminium alloy adhering to the mold. Fig. 191 shows a piston in a mold after casting, removal of cores, and opening of the mold. After removal of the casting and cleaning the mold, the molder and helper then close the mold, lock it, and replace the cores in an order reversed from that in which they were removed. This entire operation requires about 60 to 75 secs., and the piston casting actually remains in the mold for about 25 to 35 secs. after being poured. Two men can cast from 200 to 300 pistons in a 9-hour working day. On removal from the mold, the pistons are inspected while hot, and

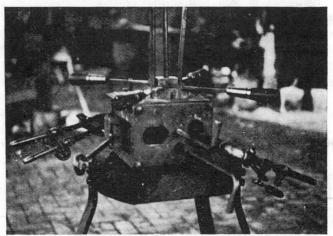

FIG. 187.—*View of mold for aluminium-alloy pistons; assembled.*

all defective castings, together with broken gates, splashings, etc., are returned to the melting furnace. After cooling, the

FIG. 188.—*Front view of mold in Fig. 187, showing the parting and lock.*

castings which have passed the hot inspection are sent to the band saw, where the gates and risers are cut off and then the castings are ground smooth, after sawing, on a grinding wheel. Each lot of pistons is then inspected cold and weighed, and then sent to a

FIG. 189.—*View of open mold with the cores removed.*

rough stock room. From the rough stock reserve the pistons are sent to the machine shop for complete machining. The

FIG. 190.—*Cores for a piston mold; assembled.*

machining operations on a typical piston are described in Chapter XIX. After machining, the castings are sent to the finished stores and held for delivery.

Ordinarily, before starting to pour pistons in permanent molds, the molds must be heated from 2 to 4 hrs., and this is usually done by lighting the mold burners at 4:00 a.m., so that the molds will be sufficiently hot when the force comes to work. The dipping ladles, and the gates and risers of the molds, are given a coating of a mixture of sodium silicate and lime to prevent the alloy from adhering. The cores must be straightened from time to time because of warping. The normal casting loss in the production of pistons in permanent molds is high, and the wasters are due principally to blowholes, shrinkage holes, cracks, and inclusions. Casting losses are not easy to control in piston production because of the large number of variables in the

FIG. 191.—*Piston in mold after casting and removal of cores.*

process. Among other factors the following variables enter into piston production, or into the production of any casting made in permanent molds, viz., (1) the composition of the raw melting stock used and the alloy made; (2) the melting temperature and period of soaking in the alloying furnace; (3) temperature of pouring; (4) method of gating, vents, feeders, etc., i.e., mechanical variables in the mold itself; (5) temperature of the mold and cores, and variation in the temperature in different parts of the mold; (6) speed of pouring; and (7) various operating irregularities, such as interrupted pouring.

Fig. 192 shows two piston castings with the gates and shrink pads attached. The question of the gating of piston castings has been studied by the author and M. E. Boyd [43] in connection with an investigation of the production of pistons in aluminium alloys for internal-combustion engines, but the details of the experimental work can not be included here.

Properties of Permanent-Mold Castings.—The mechanical properties of aluminium-alloy permanent-mold castings are essentially the same as those of otherwise identical die castings. The strength of alloys cast in permanent molds is, of course, greater than when cast in sand molds; they are more dense; the grain size is finer; and their general physical and mechanical properties are superior. Since permanent-mold castings are made in cast-iron molds, they are chilled more than sand castings. The mechanical properties of the alloys when thus cast vary with

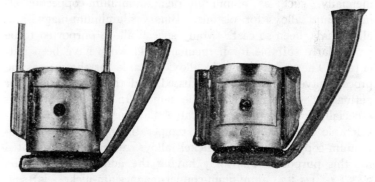

FIG. 192.—*Pistons with bottom-feed gates and two shrink pads.*

the amount of chill, and a chill-cast alloy is always stronger and more ductile than a sand-cast one. Aluminium-alloy castings when made in permanent molds are always more sound than die castings, and the minute and general porosity which is invariably found in die-cast aluminium alloys—i.e., die cast under air pressure—is not known in alloys cast in permanent molds. The comparative tensile properties of 92 : 8 aluminium-copper alloy, sand cast and chill cast, are as follows, respectively: tensile strength, 20,000 and 25,000 lb. per sq. in.; and elongation, 1.0 and 2.5 per cent. The specific gravities are 2.85 and 2.86. One of the most important features of aluminium-alloy permanent-mold castings (when properly made) is their freedom from general porosity, and such castings can be absolutely relied upon to give uniform and definite mechanical properties.

Aluminium Alloys for Permanent-Mold Castings.—All alloys that can be cast by die casting can be poured in permanent molds, but at the present time only aluminium-rich alloys and copper-rich alloys, in non-ferrous materials, are cast by the permanent-mold process. The aluminium alloys which have been used for die casting and sand casting have been cast successfully in permanent molds, and brass, bronze, and related alloys, and aluminium bronze are used for permanent-mold castings. The principal alloys employed for permanent-mold castings are the aluminium-copper alloys, exemplified by 92 : 8 aluminium-copper, and related alloys, but a number of special alloys are cast also. Certain complex aluminium alloys are cast commercially, such as aluminium-rich aluminium-copper-nickel-magnesium alloys for pistons. Binary aluminium-magnesium alloys have been so cast. Many special alloys purported to be particularly suitable for permanent-mold work have been patented. The requirements in alloys suitable for the die-casting process have already been discussed, and the requirements for permanent-mold casting are the same. In the case of alloys especially for pistons it is desirable that they should have considerable strength at elevated temperatures, and certain aluminium-copper-magnesium-nickel alloys appear well adapted for this purpose. The alloy having the nominal composition 92.5 : 4 : 1.5 : 2 aluminium-copper-magnesium-nickel is suggested as being good.

Use of Permanent-Mold Castings.—Permanent-mold castings in light aluminium alloys have not yet found general and wide application, but have been used largely as pistons for internal-combustion engines. The bulk of the present output of pistons is by permanent-mold casting, but some pistons are sand cast. Permanent-mold castings have been used in subordinate amount for the same purposes as die castings, and they are coming into use for small parts for motor cars, e.g., hardware and fittings. Camshaft bearings and valve tappett guides for motors are being cast, and steering-wheels spiders have also been made.

SELECTED BIBLIOGRAPHY.

The literature relating to die casting and permanent-mold casting, particularly the latter, is rather sparse. Many patents have been taken out appertaining to die-casting machines, various types of permanent molds for casting particular parts, and alloys purported to be especially suitable for these processes.

DIE CASTING.

1. Anon., Making dies used in the die-casting process, *The Foundry*, vol. 41, 1913, pp. 193–196.
2. Lake, E. F., Metals and alloys used for die castings, *The Foundry*, vol. 42, 1914, pp. 19–21.
3. Pack, C., Modern die-casting practice, Trans. Amer. Inst. of Metals, vol. 8, 1914, pp. 87–102.
4. Pack, C., Aluminium die casting a commercial achievement, Trans. Amer. Inst. of Metals, vol. 9, 1915, pp. 145–154.
5. Buckingham, E., Precision castings, *The Metal Ind.*, vol. 13, 1915, pp. 11–13.
6. Chapple, C. E., Die castings, *The Metal Ind.*, vol. 13, 1915, pp. 62–63.
7. Scherer, W., Die castings, *The Elec. Jour.*, vol. 12, 1915, pp. 109–112.
8. Schultz, J. A., Data on die-castings, *The Mech. World*, vol. 60, 1916, p. 28; and abstd. *Jour. Inst. of Metals*, vol. 16, 1916, p. 280.
9. Lasner, D., Die casting, *The Metal Ind.*, vol. 15, 1917, p. 252.
10. Williams, H. M., The swelling of zinc base die castings, Trans. Amer. Inst. of Metals, vol. 11, 1917, pp. 221–225.
11. Pack, C., The application of die castings in aircraft, *Aviation*, vol. 4, 1918, pp. 208–209.
12. Stern, M., The design of die castings, *Amer. Mach.*, vol. 49, 1918, pp. 549–551.
13. Hiscox, W. J., Die castings as a commercial proposition, *The Metal Ind.* (London), vol. 15, 1919, p. 408–409.
14. Oberg, E., Die casting and die-casting metals, *Machy.*, vol. 26, 1919, pp. 155–160.
15. Pack, C., Die castings and their application to the war program, Trans. Amer. Inst. of Min. Engrs., vol. 60, 1919, pp. 577–586.
16. Skinner, H. C., The manufacture and use of die-cast engine bearings, *Autom. Ind.*, vol. 42, 1920, pp. 1160–1162.
17. Pack, C., Die castings, paper before the Amer. Soc. of Mech. Engrs., St. Louis meeting, May 26, 1920.
18. Pack, C., The relation of die casting to foundry practice, Trans. Amer. Foundrymen's Assoc., vol. 29, 1921, pp. 491–498.
19. Stern, M., Application of inserts to die castings, *Amer. Mach.*, vol. 54, 1921, pp. 284–286.
20. Anon., Die-casting, *The Industrial Press*, New York, 1921.
21. Carman, A. G., Equipment for making die castings, *Machy.*, vol. 29, No. 4, 1922, pp. 261–263.
22. Pack, C., Developments in die-casting practice, *Machy.*, vol. 29, No. 4, 1922, pp. 281–283.
23. Harriman, J. W., Die-casting processes and equipment, *Amer. Mach.*, vol. 58, 1923, pp. 137–141.
24. Carman, A. G., Dies for die castings, *Machy.*, vol. 29, No. 5, 1923, pp. 369–370.
25. Carman, A. G., Dies for die castings, *Machy.*, vol. 29, No. 6, 1923, pp. 430–432.
26. Carman, A. G., Metals used for die casting, *Machy.*, vol. 29, No. 7, 1923, pp. 516–518.
27. Tour, S., Die castings, *Jour. Ind. and Eng. Chem.*, vol. 15, 1923, pp. 25–28.

PERMANENT-MOLD CASTING.

28. Anon., The Székely process of casting in metallic molds, *Engineering*, vol. 85,. 1908, pp. 16–17.

29. May, W. J., Making permanent iron molds for castings, *Mech. World*, vol. 58, 1915, pp. 91–92.
30. Barrett, H. G., Method of venting chill molds, paper before British Foundrymen's Assoc., April 23, 1915; abst., *Foundry Trade Jour.*, vol. for 1915, pp. 309–310.
31. Johnson, F., Casting of non-ferrous alloys in chill molds, Proc. British Foundrymen's Assoc., vol. for 1915–1916, pp. 148–162.
32. Johnson, B. W., The prehistoric founder, Proc. British Foundrymen's Assoc., vol. for 1915–1916, pp. 249–298.
33. Rix, H., and Whitaker, H., Die casting of aluminium bronze, *Jour. Inst. of Metals*, vol. 19, 1918, pp. 123–131.
34. Anon., Aluminium chill and die castings, *The Metal Ind.* (London), vol. 16, 1920, pp. 431–432.
35. Anon., Dies for hand poured die castings, *The Metal Ind.* (London), vol. 19, 1921, pp. 41–44; 97–99; 128–131.
36. Anon., An outline and comparison of the die-casting processes, *The Metal Ind.* (London), vol. 18, 1921, pp. 481–483.
37. Guillet, L., Description de quelques coquilles de moulage utilisées pour l'aluminium et ses alliages, *Rev. de Mét.*, vol. 18, 1921, pp. 512–516.
38. Wells, S. A. E., Casting in metal molds, *The Metal Ind.* (London), vol. 19, 1921, pp. 501–502.
39. Bertoya, O., Permanent mold castings, *Engng. Prod.*, vol. 2, 1921, pp. 382–383.
40. Diamond, J. E., The aluminium-alloy piston, *Jour. Soc. Autom. Engrs.*, vol. 11, 1922, pp. 258–261.
41. Anon., Some experiences of aluminium and its alloys for motor pistons, *The Metal Ind.* (London), vol. 20, 1922, p. 321.
42. Anderson, R. J., and Boyd, M. E., The production of castings in permanent molds, paper before the Inst. of British Foundrymen, Newcastle-on-Tyne meeting, June, 1924.
43. Anderson, R. J., and Boyd, M. E., The production of aluminium-alloy pistons in permanent molds, paper before the Amer. Foundrymen's Assoc., Milwaukee meeting, Oct., 1924.